30-SECOND
BRAIN

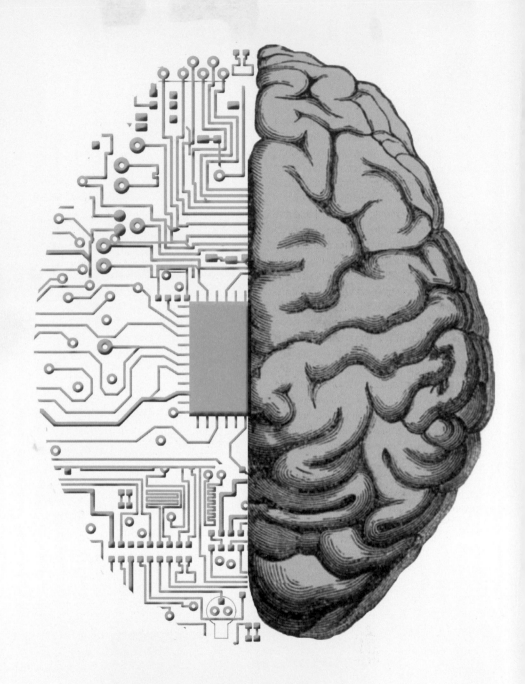

30-SECOND
BRAIN

THE 50 MOST MIND-BLOWING IDEAS IN NEUROSCIENCE,
EACH EXPLAINED IN HALF A MINUTE

Editor
Anil Seth

Foreword by
Chris Frith

Contributors
Tristan Bekinschtein
Daniel Bor
Christian Jarrett
Ryota Kanai
Michael O'Shea
Anil Seth
Jamie Ward

Illustrations
Ivan Hissey

ICON

This edition published in the UK in 2018 by
Icon Books Ltd
Omnibus Business Centre
39–41 North Road, London N7 9DP
email: info@iconbooks.com
www.iconbooks.com

First published in the UK in 2013

This book was conceived,
designed and produced by
Ivy Press
An imprint of The Quarto Group
The Old Brewery
6 Blundell Street
London N7 9BH, UK
T (0)20 7700 6700
F (0)20 7700 8066
www.QuartoKnows.com

Creative Director **Peter Bridgewater**
Publisher **Jason Hook**
Editorial Director **Caroline Earle**
Art Director **Michael Whitehead**
Designer **Ginny Zeal**
Picture Research **Jamie Pumfrey**
Glossaries Text **Anil Seth**
Profiles Text **Viv Croot**

ISBN: 978-1-78578-356-2

Printed in China

10 9 8 7 6 5 4 3 2 1

MIX
Paper from
responsible sources
FSC® C017606
FSC
www.fsc.org

CONTENTS

FOREWORD
Chris Frith

The human brain is the most complex entity we
know of. It contains at least 90 billion neurons (nerve cells). Each of these
is a complex information-processing device in its own right and interacts
with about 1,000 other neurons. Understanding this degree of complexity
is a daunting task.

Our understanding of the human brain is still in its infancy. The identification
of the neuron as the basic building block of the brain occurred only a hundred
years ago. At first, progress depended on the study of damaged brains. It
is only in the last twenty-five years that it has become possible to see brain
structure and function in healthy volunteers. The remarkably detailed images
that emerge from brain scanners, with their brightly coloured blobs, have
had a dramatic impact. Human brains have become the image of choice
for the media, attached to articles about 'What our brains can teach us'
or 'Contours of the mind'.

Brain research is beginning to attract big money. The Brain Activity Map
project is expected to receive $3 billion from the US Government over the
next ten years. The hope is that investigating the human brain in exquisite
detail will have a similar pay-off to that achieved by the human genome
project and will lead to progress in understanding mental disorders,
such as autism and schizophrenia.

One of the most exciting features of research on the human brain
is that we confront deep philosophical questions. Minds depend on brains.
Without brains we could not think, or feel, or imagine. But we still feel
uncomfortable with this identity. Am I simply the product of electrical activity
in my brain? How can subjective experience emerge from brain activity?

Our theories about how brains work remain very primitive. Some people
think that an insoluble conundrum arises because the human brain is trying
to understand itself. Surely something complex can only be understood by
something even more complex? I believe that this problem is more apparent
than real. Here's why. One of the glories of the human brain is that it
enables us to share our thoughts. Our understanding is built on the thinking

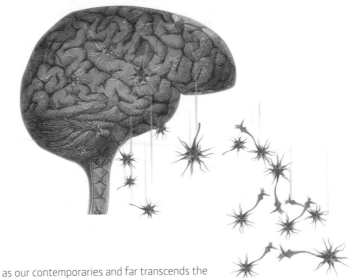

Sophisticated network
The brain is fired up by a network of 90 billion neurons. Neuroscientists are only just beginning to discover how their activity relates to what happens in our minds.

of our predecessors as well as our contemporaries and far transcends the abilities of any single brain. We pay too little attention to these effects of culture and collaboration.

Consider another highly social animal, the bee. The brain of a bee weighs 1 milligram and contains a mere million neurons. Yet this tiny brain enables bees to learn about the world and communicate using their waggle dance. Even more impressive is what bees can achieve through collaboration. From the reports of scouts, a swarm of bees can make a group decisions about the best site for a new nest.

Recent studies suggest that the way the bees interact to make decisions closely resembles the way neurons in the human brain interact to make decisions. This comparison gives us a feel for the dramatically enhanced abilities of the human brain compared with the bee brain. But it also instils in me a sense of wonder about what humans can achieve as a group.

A group of bees working together can achieve abilities resembling those of a single human brain. Imagine an entity containing the power of multiple, interacting human brains. We create such a system whenever we interact. And the best example of the power of such a system comes from the practice of science. It is through the practice of science that we will be able to unravel the mysteries of the brain. This book shows how exciting this journey will be.

INTRODUCTION
Anil Seth

Understanding how the brain works is one of our greatest scientific quests. The challenge is quite different from other frontiers in science. Unlike the bizarre world of the very small in which quantum-mechanical particles can exist and not exist at the same time, or the mind-boggling expanses of time and space conjured up in astronomy, the human brain is in one sense an everyday object; it is about the size and shape of a cauliflower, weighs about 1.36 kg (3 lb) and has a texture like tofu. It is the complexity of the brain that makes it so remarkable and difficult to fathom. There are so many connections in the average adult human brain that if you counted one each second it would take you more than 3 million years to finish.

Faced with such a daunting prospect it might seem as well to give up and do some gardening instead. But the brain cannot be ignored. As we live longer, more and more of us are experiencing – or will experience – neurodegenerative conditions, such as Alzheimer's disease. The incidence of psychiatric illnesses, such as depression and schizophrenia, is also on the rise. Better treatments for these conditions depend on a better understanding of the brain's intricate networks.

More fundamentally, the brain draws us in because it defines who we are. It is much more than just a machine to think with. Hippocrates, the father of Western medicine, recognized this long ago: 'Men ought to know that from nothing else but the brain come joys, delights, laughter and jests, and sorrows, griefs, despondency and lamentations.' More recently Francis Crick – one of the major biologists of our time (see the biography on page 80) – echoed the same idea: 'You, your joys and your sorrows, your memories and your ambitions, your sense of personal identity and free will, are in fact no more than the behaviour of a vast assembly of nerve cells and their associated molecules.' And, perhaps less controversially, but just as important, the brain is also responsible for the way we perceive the world and how we behave within it. So to understand the brain is to understand our own selves and our place in society and in nature.

More than a machine
The brain is a complex and intricate information-processing mechanism – not just for cold, hard facts but for how we move, feel, laugh and cry. Neuroscientists are constantly gaining new insights into the inner workings of the brain.

But how to begin? From humble beginnings, neuroscience is now a vast enterprise involving scientists from many different disciplines and almost every country in the world. The annual meeting of the Society for Neuroscience attracts more than 20,000 (and sometimes more than 30,000) brain scientists. No single person – however capacious their brain – could possibly keep track of such an enormous and fast-moving field. Fortunately, as in any area of science, underlying all this complexity are some key ideas to help us get by. Here's where this book can help.

How the book works

Within the following pages, leading neuroscientists and science writers will take you on a tour of 50 of the most exciting ideas in modern brain science, using simple plain English. To start with, in **Building the Brain** we will learn about the basic components and design of the brain, and trace its history from birth (and before), and through evolution. **Brain Theories** will introduce some of the most promising ideas about how the brain's many billions of nerve cells (neurons) work together. **Mapping the Brain** will show how new technologies are enabling us to chart the brain's intricate structure and activity patterns. Then, in **Consciousness**, we tackle the still mysterious relationship between the brain and conscious experience – how does the buzzing of neurons transform into the subjective experience of being you, here, now, reading these words? In the following chapters, **Perception & Action** and **Cognition & Emotion**, we will explore how the brain enables these important functions, both with and without consciousness. Finally, in the last chapter – **The Changing Brain** – we will explore some recent ideas about how the brain changes its structure and function throughout life in both health and in disease.

Approach the book however you like. Read it in order, or dip in and out. Each of the 50 ideas is condensed into a concise, accessible and engaging '30-second neuroscience'. To get the main message across, there is also a '3-second brainwave', and a '3-minute brainstorm' provides some extra food for thought on each topic. There are helpful glossaries summarizing the most important terms used in each chapter, as well as biographies of key scientists who helped make neuroscience what it is today. Above all, I hope to convey that the science of the brain is just getting into its stride. These are exciting times and it's time to put the old grey matter through its paces.

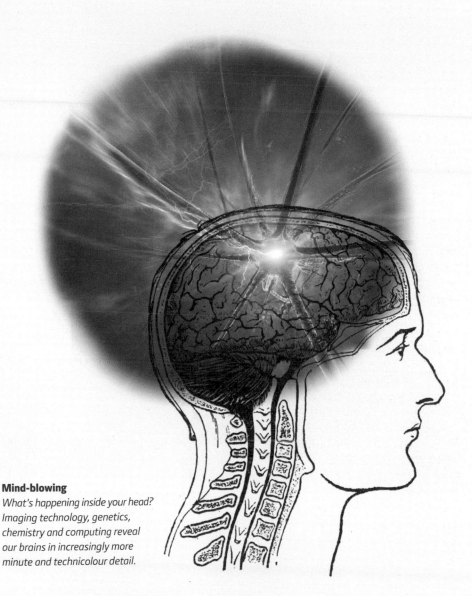

Mind-blowing

What's happening inside your head? Imaging technology, genetics, chemistry and computing reveal our brains in increasingly more minute and technicolour detail.

BUILDING THE BRAIN

axon A long, thin fibre extending from the cell body (soma) of a neuron, conveying its output in the form of a spike (nerve impulse or action potential) and enabling communication with other neurons. Each neuron will have at most one axon. Axons typically split into many separate branches before connecting with the dendrites of other neurons.

brain stem A small stalk-like area at the bottom of the brain, lying in between the spinal cord and the rest of the brain. The brain stem controls many vital basic bodily functions, such as breathing, swallowing and blood pressure regulation. Because so many neural pathways pass through the brain stem, damage to this area can have profound effects.

cerebral cortex The deeply folded outer layers of the brain, which take up about two-thirds of its entire volume and are divided into left and right hemispheres that house the majority of the 'grey matter' (so called because of the lack of myelination that makes other parts of the brain seem white). The cerebral cortex is separated into lobes, each having different functions, including perception, thought, language, action and other 'higher' cognitive processes, such as decision making.

dendrites The short input fibres of a neuron that are organized into complicated tree-like patterns. Each neuron has many dendrites that make contact with axons from other neurons via synapses. Dendrites convey the incoming signals to the cell body (soma) of a neuron, which will then produce an output of its own.

frontal lobes One of the four main divisions of the cerebral cortex and the most highly developed in humans compared with other animals. The frontal lobes (one for each hemisphere) house areas associated with decision making, planning, memory, voluntary action and personality.

hippocampus A seahorse-shaped area found deep within the temporal lobes. The hippocampus is associated with the formation and consolidation of memories and also supports spatial navigation. Damage to this area can lead to severe amnesia, especially for episodic (autobiographical) memories.

myelination A process by which a neuron's axons are coated with myelin, which both insulates the axon from other nearby axons and dramatically increases the speed of nerve impulses (spikes) travelling along it. Myelination, which relies on glial cells, is essential for efficient transmission of information in the brain.

occipital lobes Another of the four main divisions of the cerebral cortex, the occipital lobes are at the back of the brain and house regions mainly involved in vision. Damage to the occipital lobes can result in blindness or more selective deficits.

olfactory system One of the most evolutionarily ancient parts of the brain. The olfactory system underpins the sense of smell and is less well-developed in humans than in many other animals. Signals from olfactory sensory neurons in the nose are conveyed to the olfactory bulb deep inside the brain. Olfaction and taste are distinct from the other senses by responding to chemical stimulation.

parietal lobes The third major division of the cerebral cortex. The parietal lobes lie above the occipital lobes and behind the frontal lobes and are deeply involved in integrating information from the different senses. The parietal cortex is essential for organizing our experience of space and position and it is heavily involved in attentional processes.

Purkinje cells Found exclusively in the cerebellum, these neurons are among the largest in the brain and have elaborately branching dendritic structures. Purkinje cells provide long-range inhibitory control over output parts of the cerebellum, enabling fine motor coordination and error correction.

synapses The junctions between neurons, linking the axon of one to a dendrite of another. Synapses ensure that neurons are physically separate from each other so that the brain is not one continuous mesh. Communication across synapses can happen either chemically via neurotransmitters or electrically.

temporal lobes The last of the four main divisions of the cerebral cortex. These lobes are found low to the side of each hemisphere and are heavily involved in object recognition, memory formation and storage and language. The hippocampus is in the medial part of these lobes (the medial temporal lobe).

thalamus These are bundles (nuclei) of neurons that sit on top of the brain stem and are about the size and shape of a walnut. The thalamic nuclei are heavily interconnected with specific areas of the cerebral cortex and are thought to act as sensory relay areas, connecting sensory receptors (apart from olfaction) with the cortex.

NEURONS & GLIAL CELLS

the 30-second neuroscience

3-SECOND BRAINWAVE
There are 4 km (2½ miles)
of neuronal network
interconnections packed
into every cubic millimetre
of grey matter.

3-MINUTE BRAINSTORM
Could you think yourself
thin? The brain is just two
per cent of your body
weight but consumes
twenty per cent of your
daily energy needs.
Exercising the brain is
energetically expensive.
In spite of this, as humans
evolved, the most
thoughtful part of the
cerebral cortex rapidly
tripled in size beginning
about two million years ago.
Most of the additional cost
of evolving our uniquely
human cognitive abilities
is consumed by a single
enzyme that recharges the
batteries that power
electrical nerve impulses.

Your neurons (your marbles,
if you prefer) are the information processing cells
of your brain. You have between 90 and 100 billion
of them, yet not one of them has any idea who
you are. But somehow, by chattering among
themselves across networks of billions of
interconnections, neurons conjure up your
self-awareness. Neurons receive messages from
other neurons on their cell body and its short
extensions – called dendrites – at specialized
structures called synapses. Messages are sent
to other neurons via long, slender fibres – called
axons – in coded patterns of electrical spikes
(nerve impulses). Each impulse is about 0.1 volt
and lasts one- to two-thousandths of a second,
hurtling along axons at up to 480 kph (300 mph).
Arriving at a synapse, impulses trigger the release
of signalling chemicals called neurotransmitters.
These alter the pattern of spikes generated
by the receiving neuron. And that is basically
how the brain works. Well, not quite. Neurons
work properly only if bathed in the right blend
of chemicals. Glial cells, which outnumber
neurons 50:1, maintain this condition. They help
neurons wire together in the developing brain,
nurture them in the adult brain, insulate axons,
mop up dead cells, recycle used neurotransmitters
and protect the brain from infection. They are the
unsung heroes of the brain's story.

RELATED BRAINPOWER
See also
NEUROTRANSMITTERS
& RECEPTORS
page 18

NEURAL NETWORKS
page 40

3-SECOND BIOGRAPHIES
SANTIAGO RAMÓN Y CAJAL
1852–1934
Anatomist who defined the
cellular components of mental
activity

WALTHER NERNST
1864–1941
His theoretical work explained
how voltages are generated
by cells

BERNARD KATZ
1911–2003
Proposed the quantum/vesicular
hypothesis of neurotransmitter
release

30-SECOND TEXT
Michael O'Shea

*For every networking
executive neuron, there
are 50 low-status but
essential glial cells
maintaining the neural
environment.*

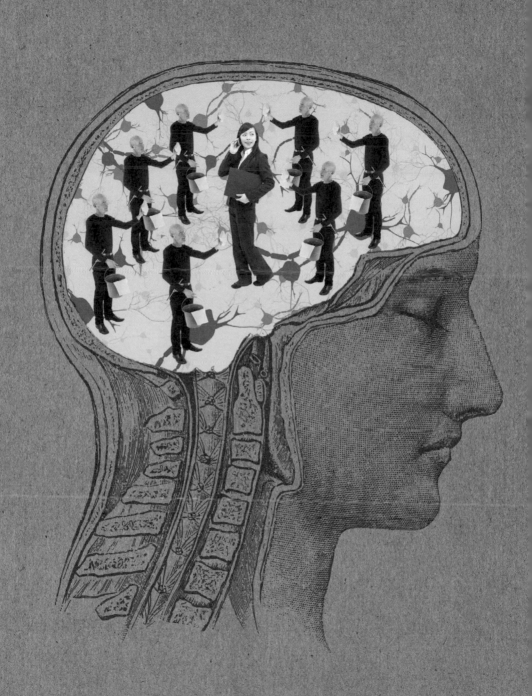

NEUROTRANSMITTERS & RECEPTORS

the 30-second neuroscience

3-SECOND BRAINWAVE
Active neurons release
neurotransmitters that
activate receptors in
other neurons to change
information flow in the
brain in the short, medium
and long term.

3-MINUTE BRAINSTORM
Nitric oxide (NO), a
poisonous gas, is a most
unlikely neurotransmitter.
It cannot be stored in
vesicles, so it is released
as it is produced, within
specialized active neurons.
NO then spreads into
swathes of the brain, where
it can affect many receptive
neurons without the
transmitting neuron having
to be directly connected to
them. This 'non-synaptic'
signalling is important in
long-term memory
formation.

Neurotransmitters convey signals
between neurons, briefly exciting or inhibiting
their electrical activity. They are released when
nerve impulses arrive at synapses. They range
from very small molecules, to medium compounds,
to giant molecules called peptides. They are
stored in tiny spheres called synaptic vesicles.
Impulses cause the vesicles to release their
contents into the synaptic gap between
transmitting and receiving neurons. Released
neurotransmitters act by binding to receptor
proteins, each of which is tuned to just one
neurotransmitter type. There are scores of
neurotransmitters and even more receptors.
Why so many? After all, if neurotransmitters
mediate just two simple functions – excitation
and inhibition – surely two transmitters and their
receptors is enough? Things are not so simple.
Many neurotransmitters do not trigger fast
excitation or inhibition, but initiate quite slow
metabolic processes in neurons, causing lasting
changes in the strength of synaptic connections.
Neurotransmitters can also initiate the switching
ON and OFF of important genes, which can
cause long-term change in neuronal and synaptic
properties. Are these the changes in the brain on
which memories depend? Probably, but we are far
from a complete understanding of the brain's
complex chemical language.

RELATED BRAINPOWER
See also
NEURONS & GLIAL CELLS
page 16

3-SECOND BIOGRAPHIES
OTTO LOEWI
1873–1961
First to show that a stimulated
nerve releases a substance that
has a physiological effect

HENRY DALE
1875–1968
Most famous for the so- called
Dale's Principle – that all synapses
of a single neuron release the
same neurotransmitter(s)

BERNARD KATZ
1911–2003
Proposed the quantum/vesicular
hypothesis of neurotransmitter
release

30-SECOND TEXT
Michael O'Shea

*While you were
pondering whether to
order pizza or tacos,
a sophisticated
performance of
chemical signalling
just took place in
your brain.*

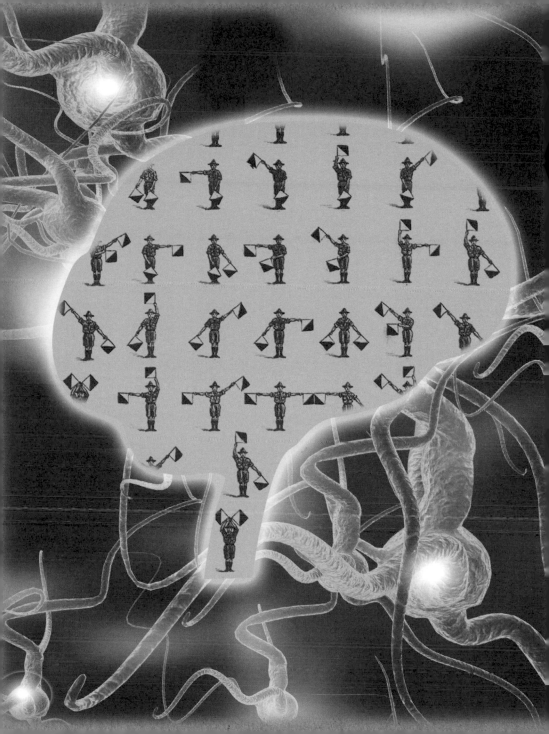

NEUROGENETICS

the 30-second neuroscience

3-SECOND BRAINWAVE
The brain uses 70 per cent of our 22,000 genes. Those affecting synaptic function are particularly important because their activity can be regulated by experience.

3-MINUTE BRAINSTORM
Genes have an important role in disorders of the mind and behaviour, such as ADHD, autism, bipolar disorder, depression and schizophrenia. In fact, while these are considered as clinically different, recent research suggests that they share genetic risk factors. The identification of shared genetic causes of a range of psychiatric disorders may lead to the discovery of an underlying molecular mechanism for mental illness. This would represent a major advance in the development of preventative medicines.

A gene is a set of instructions

in DNA for making a protein. There are about 22,000 genes in the human genome. Although proteins are the essential cogs and levers in the functioning of all neurons, no cell needs all 22,000 genes. So neurons, along with other cells, turn on only the genes required for their own needs. As needs change, different genes are turned on or off. This changing pattern of active genes is particularly notable in the functioning of synapses. This is important because changing the connections in neural circuits allows us to learn from experience. Consider a neural circuit that detects a potentially threatening sensory stimulus. If the threat persists, strengthened circuit connections will be required to sustain and enhance vigilance. To achieve this, signals are dispatched from the sharp end of neurons – the synapses – to their central nuclei and there the DNA is ordered to turn on the required genes. Freshly made synapse-reinforcing proteins are then rushed back to the same synapses that ordered them. So while genes certainly affect brain function, the fact that they can be influenced by their environment frees our behaviour from a rigid genetic determinism as the brain's genetic machinery responds adaptively to changing circumstances.

RELATED BRAINPOWER
See also
NEURONS & GLIAL CELLS
page 16

THE EVOLVING BRAIN
page 30

THE SCHIZOPHRENIC BRAIN
page 150

3-SECOND BIOGRAPHIES
FRANCIS CRICK &
JAMES WATSON
1916–2004 & 1928–
Awarded the Nobel Prize in 1962 for determining the structure of DNA and suggesting how DNA encodes and replicates genetic information

30-SECOND TEXT
Michael O'Shea

Neurons turn genes on or off to order protein material when needed; it's a little like just-in-time delivery systems, but it works.

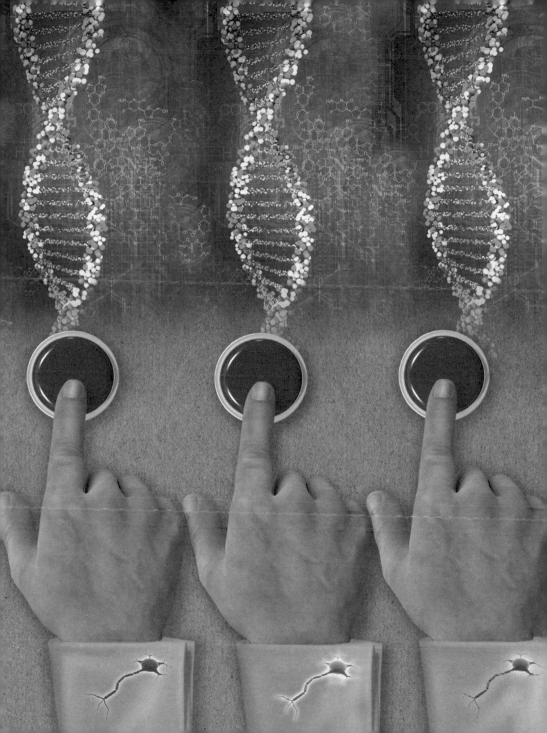

1 May 1852
Born in Petilla de
Aragón, Spain

1873
Graduated from the
medical school of the
University of Zaragoza

1874–75
Served as an army
doctor, accompanied
expedition to Cuba

1883
Appointed Chair
of Anatomy at the
University of Valencia

1888–94
Published *Revista
Trimestral de Histología
Normal y Patológica*,
results of his systematic
histological study of the
nervous system

1888
Discovered that axons
terminate freely and the
existence of dendritic
spines on neural
dendrites

1891
Promulgated his theory
of the individuality of
the nerve cell

1892
Published his Law of
Dynamic Polarization

1901
Appointed director of
the Biological Research
Laboratory, which would
become the Cajal
Institute in 1922

1906
Shared Nobel Prize for
Physiology or Medicine
with Camillo Golgi for
their work on the
structure of the
nervous system

17 October 1934
Died in Madrid

SANTIAGO RAMÓN Y CAJAL

Acknowledged by many to be

the architect of modern neurobiology, as a young man Cajal tried very hard to keep out of medicine altogether. He wanted to be an artist, but his father (a professor of dissection) was equally adamant that he should be a doctor. After miserable but educational apprenticeships with a cobbler and a barber, Cajal gained his medical licence and buckled down to join the family business, eventually being appointed Professor of Anatomy at Valencia.

He kept up with his drawing, making many anatomical studies, and it is possible that it was his artistic eye that led him to his greatest discovery. When in 1887, now at the University of Barcelona, he looked at the Italian physician Camillo Golgi's immaculately stained slides of brain cells, he saw what others had not. Until this time, the prevailing orthodoxy was that the nervous system was a single reticular (mesh-like) construct without discrete cellular components (neurons). Cajal realized however that what Golgi's images clearly showed was that the nervous system was a network of discrete cellular components. This was a correct interpretation that crucially allowed neurons to be regarded as the functional units of the brain – free agents that could form many synaptic connections, each

capable of being modified to allow for growth and adaptation. Cajal studied this newly revealed phenomenon for four years, identifying as well so-called 'dendritic spines' – small membranous protrusions from a neuron's input fibres that typically receive input from a single synapse. He used his artistic skills to make meticulous drawings and then published his findings in his magnum opus *Revista Trimestral de Histología Normal y Patológica*, which had an impact on the scientific community similar to Darwin's breakthrough *On the Origin of Species*.

By providing the most accurate description of the neuron's function and mechanism, it changed the way neuroscience worked and cleared the path for the formulation of neuron doctrine proposed by German anatomist Heinrich von Waldeyer-Hartz. Cajal was a prodigious publisher and contributor to medical journals. He was greatly feted and heaped with awards, including the 1906 Nobel Prize for Physiology or Medicine, which he shared with Golgi. Cajal also found time to work in other areas of medicine, notably cancer, and to set up his own research institute in Madrid. For all the breadth of his achievements, he will always be best remembered for disentangling the neuron from its imagined network.

THE BASIC ARCHITECTURE OF THE BRAIN

the 30-second neuroscience

3-SECOND BRAINWAVE
The brain can be divided crudely into three basic parts: the outer cortex; the diencephalon, including the thalamus; and the brain stem.

3-MINUTE BRAINSTORM
We now take it for granted that the brain is the seat of thought, but this wasn't always the case. Even after the importance of the brain was demonstrated by Galen in the second century CE, it would take more than a millennium for this view to be universally accepted. Writing as late as the seventeenth century, the English philosopher Henry More argued the human brain has as much potential for thought as 'a bowl of curds'.

Imagine you've just picked up a typical 1.36-kg (3-lb) adult human brain. The outer spongy tissue that you grip in each hand is the cortex. Look at the apparently random pattern of grooves on the surface – the sulci – and you'll see some deeper lines. These landmarks show the divisions between the main lobes of the cortex: the frontal lobes, the temporal lobes near the ears, the parietal lobe at the crown of the head and the occipital lobe at the rear. Lift the brain above your head and sprouting underneath you will see the brain stem, responsible for regulating the most basic life-sustaining functions, including breathing and heart rate. Also note the cauliflower-like cerebellum nestled next door. In a living person, the brain stem would connect to the spinal cord, thereby linking the brain with the rest of the body. Now place the brain back down and gently pry apart the two hemispheres so that you reveal the inner structures, including the top of the brain stem, known as the midbrain. Above this is the egg-shaped thalamus – the brain's relay station. Nearly all incoming sensory information is sent here before being passed on to the cortex. In the traditional language of anatomy, the brain stem is the metencephalon, the thalamus is part of the diencephalon and the outer cortex is the telencephalon.

RELATED BRAINPOWER
See also
NEURONS & GLIAL CELLS
page 16

THE CEREBELLUM
page 26

THE DEVELOPING BRAIN
page 28

THE LOCALIZATION OF FUNCTION
page 36

LEFT BRAIN VS RIGHT BRAIN
page 68

3-SECOND BIOGRAPHY
GALEN
129–CA 210/216 CE
The 'prince of physicians', credited with providing the first demonstration of the importance of the brain to behaviour

30-SECOND TEXT
Christian Jarrett

Blueprint for the brain: A compact space-saving multi-functional unit built to respond instantly to environmental changes; organically fuelled.

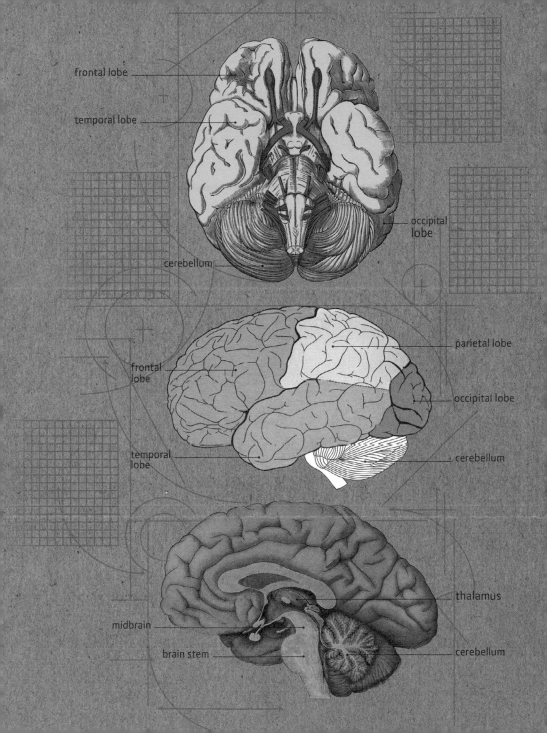

frontal lobe

temporal lobe

occipital lobe

cerebellum

parietal lobe

frontal lobe

occipital lobe

temporal lobe

cerebellum

thalamus

midbrain

brain stem

cerebellum

THE CEREBELLUM

the 30-second neuroscience

3-SECOND BRAINWAVE
The cerebellum, or 'little brain', is densely packed with neurons and its primary function is in motor control and coordination, although we now know it does much more.

3-MINUTE BRAINSTORM
The cerebellum is responsible for the fact that we are unable to tickle ourselves. As well as calculating the movements necessary to achieve a desired action (known as an 'inverse model'), another of the cerebellum's roles is to form predictions ('forward models') of the probable sensory consequences of our own actions and to cancel them out. Self-tickling doesn't work because of this process.

Hanging off the back of the brain

is a second 'little brain' (the literal translation of 'cerebellum'), resembling a fist-sized cauliflower. Densely packed with cells, it accounts for ten per cent of the brain's volume and yet contains around half the neurons found in the entire central nervous system. Like the big brain, the cerebellum is comprised of two hemispheres, except here they are joined by a narrow structure known as the vermis (literally 'worm'). Further, in common with the cerebral cortex, the highly convoluted cerebellar cortex is made up of white matter in its deeper parts, with grey matter nearer the surface. The cerebellum contains many intricately branched Purkinje cells, which are found only in this brain structure. Since at least the early nineteenth century, neuroscientists have recognized the important role played by the cerebellum in the control of movement and posture. Abnormalities in its function, whether caused by inherited disease, brain damage or the effect of alcohol, lead to difficulties in walking and a general clumsiness of movement. In recent years, our understanding of the cerebellum has undergone a revolution and it is now thought to be involved not just in motor control but in memory, mood, language and attention.

RELATED BRAINPOWER
See also
THE BASIC ARCHITECTURE
OF THE BRAIN
page 24

HOW WE PICK UP A CUP
OF COFFEE
page 108

3-SECOND BIOGRAPHIES
JAN EVANGELISTA PURKINJE
1787–1869
Described the intricately branching Purkinje cells that now bear his name

SANTIAGO RAMÓN Y CAJAL
1852–1934
Used revolutionary techniques to reveal the underlying cellular structure of the cerebellum

MASAO ITO
1928–
Pioneer in characterizing the functional circuits of the cerebellum

30-SECOND TEXT
Christian Jarrett

Unable to unicycle blindfold across a tightrope? Blame it on your cerebellum.

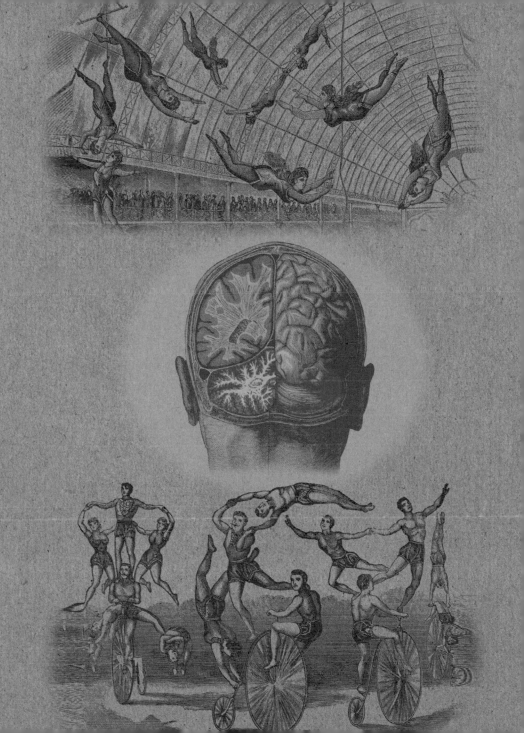

THE DEVELOPING BRAIN

the 30-second neuroscience

3-SECOND BRAINWAVE
The properties of the brain are maintained by dynamic, plastic mechanisms that originate in the embryo but continue to operate after birth and on into adulthood.

3-MINUTE BRAINSTORM
Brain development is characterized by a convergence of genetic (nature) and environmental (nurture) influences. A misunderstanding of the interaction between them lies behind questions such as: *Is this or that trait due to nature or nurture?* But it is wrong to present the question as an 'either/or' proposition. The genome does not contain enough information to make a brain on its own, so genes have evolved to exploit information coming from the environment; information essential for fine-tuning developing neuronal networks.

The brain develops from a hollow tube formed from the skin of the very early embryo. Cells multiply more rapidly at the front of the tube, which enlarges, becoming the embryonic brain. Newly produced cells transform into immature neurons. By the time the embryo is about four weeks old, these migrate to their destinations, growing dendrites and axons and forming the first of what will be trillions of synaptic connections. There is no blueprint for these interconnections – the embryonic brain generates an excess of neurons and synapses, allowing competition and interactions with the environment to sculpt functional circuits. About half the embryonic neurons are killed off, having failed to form useful connections. Some surviving neurons – involved in transmitting information over long distances – have their axons insulated by glial cells, a process called myelination, which increases the speed and quality of information transmission. Until recently, brain development was thought to be completed in early childhood. In fact, grey matter volume increases gradually through childhood, peaks in early adolescence and shrinks as an adolescent becomes an adult. This reduction in brain volume seems odd, but reflects the brain's ability to adapt to the environment by pruning unused synapses and strengthening useful ones (or so we may tell ourselves!).

RELATED BRAINPOWER
See also
NEURONS & GLIAL CELLS
page 16

NEURAL DARWINISM
page 48

NEUROGENESIS & NEUROPLASTICITY
page 138

3-SECOND BIOGRAPHIES
RITA LEVI-MONTALCINI
1909–2012
Won Nobel Prize in 1986 for discovering 'nerve growth factor' (with Stanley Cohen), a key chemical shaping neural development

ROGER SPERRY
1913–94
Won Nobel Prize in 1981; he showed that chemical signals provide the basic mechanisms for wiring the brain

30-SECOND TEXT
Michael O'Shea

You're not losing brain cells, you are culling the slackers to reveal a diamond-precision, lean, mean thinking machine.

THE EVOLVING BRAIN

the 30-second neuroscience

3-SECOND BRAINWAVE
Brains began evolving millions of years ago, allowing organisms to search and respond to the external world in ever more sophisticated and flexible ways.

3-MINUTE BRAINSTORM
A recurring debate is whether the human brain is continuing to change. Genetic evidence published in 2005 suggested that it is. A team at the University of Chicago identified two versions of genes involved in brain development that had appeared relatively recently in human history – microcephalin and ASPM. The first appeared around 37,000 years ago, the other approximately 5,800 years ago. Their rapid and continuing spread through the population suggests they confer some kind of advantage.

The origins of brains can be found about a billion years ago with the appearance of the first multicellular organisms. Their cells needed to communicate with each other, so they evolved neural nets – a kind of diffuse proto-brain still found in some creatures today, such as jellyfish. Later geological and climatic events provided new environments and challenges that spurred further brain evolution, including the emergence of groups of neurons specialized for specific tasks. It's tricky to pin down when these neuronal hubs connected together to form the first brain, but we know that around half a billion years ago the fish-like ancestors of modern-day vertebrates had brain-like structures. Looking at the animal kingdom today, we can see how evolutionary pressures shaped the emergence of different types of brain. The fruit fly, for instance, lacks a cortex but has large antennal lobes and 'mushroom bodies' dedicated to processing smell. The rat has large areas of cortex devoted to processing information from its whiskers. Fish have an enlarged cerebellum, a structure involved in movement. There are many theories for what caused the massive expansion of the human brain, including bipedalism (which freed up the hands for tool use), larger social groups and the emergence of language.

RELATED BRAINPOWER
See also
NEURONS & GLIAL CELLS
page 16

NEUROGENETICS
page 20

THE CEREBELLUM
page 26

THE DEVELOPING BRAIN
page 28

30-SECOND TEXT
Christian Jarrett

The human brain is in a constant state of upgrade to cope with standing upright, opposable thumbs and the insatiable urge to text.

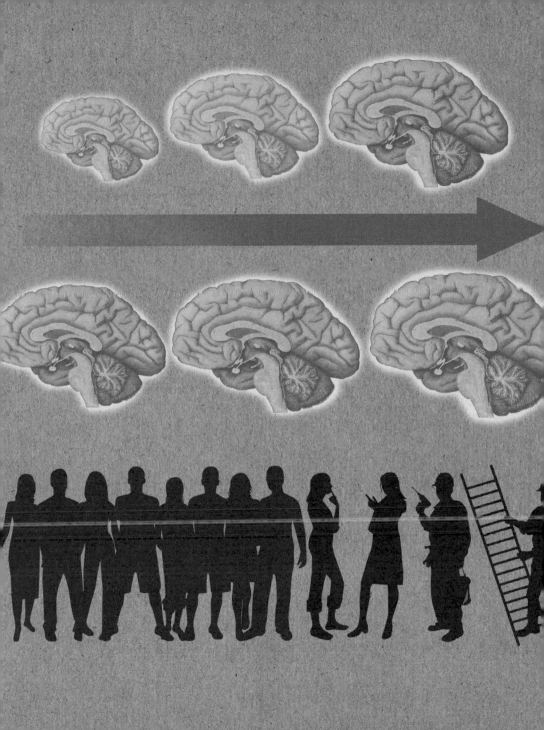

BRAIN THEORIES

Bayes' theorem A pillar of probability theory that provides a way of updating beliefs with new evidence. Named after the Reverend Thomas Bayes, Bayes' theorem relates the probability of a belief given some evidence to the prior probabilities of the belief and the evidence, and the probability of the evidence given the belief. Because it depends on prior beliefs, Bayes' theorem has endured a great deal of controversy, but it is now a mainstay within statistics and is increasingly influential as a metaphor or mechanism for how the brain might work.

cerebellum The cauliflower-shaped 'little brain' at the bottom of the main brain. It is most often associated with controlling the precision, accuracy and fluency of movements but is now also known to play important roles in other cognitive processes. Surprisingly, the cerebellum contains more neurons than the rest of the brain.

conditioning A learning process by which an event becomes associated with an outcome leading to a change in behaviour. There are many kinds of conditioning – the best known are 'classical' and 'instrumental' (or 'operant'). In classical conditioning, an environmental stimulus (such as a high-pitched sound) becomes associated with an outcome (such as the presentation of food) leading to a new behaviour (such as salivation and/or approach) with respect to the sound. In instrumental conditioning, an association is established between an action (such as pressing a lever) and an outcome (such as getting food) leading to an increased propensity to press the lever.

connectionism A theoretical approach that emphasizes how interconnected networks of neurons can learn via simple local rules applied between pairs of neurons. Connectionism is both a model of brain function and a method within artificial intelligence.

hypersynchrony A hypersynchronous state in the brain occurs when levels of neuronal synchrony exceed normal limits, so that large parts of the brain start switching on and off together. These 'electrical storms' are associated with epileptic seizures.

neurons The cellular building blocks of the brain. Neurons carry out the brain's basic operations, taking inputs from other neurons via dendrites, and – depending on the pattern or strength of these inputs – either releasing or not releasing a nerve impulse as an output. Neurons come in different varieties but (almost) all have dendrites, a cell body (soma) and a single axon.

phrenology A now discredited theory proposing that individual variation in mental capabilities and personality characteristics can be inferred on the basis of differences in the shape of the skull.

predictive coding A popular implementation of the Bayesian Brain hypothesis, according to which the brain maintains predictive models of the external causes of sensory inputs and updates these models according to some version of Bayes' theorem. Predictive coding has its roots in the ideas of Hermann von Helmholtz, who conceived of perception as a form of inference.

re-entry In terms of brain structure, re-entry describes a pattern of connections in which an area A connects to an area B, and B is reciprocally connected back to A. In terms of dynamics, re-entrant connectivity implies that neural signals flow in both directions between two areas. Re-entry should be distinguished from the term 'feedback', which is usually used to describe the processing of error signals.

synapses The junctions between neurons, linking the axon of one to a dendrite of another. Synapses ensure that neurons are physically separate from each other so that the brain is not one continuous mesh. Communication across synapses can happen either chemically via neurotransmitters, or electrically.

synchrony Applied to neuroscience, synchrony describes the correlated activity of individual neurons. Neurons in synchrony fire spikes (action potentials) at the same time, which can increase their impact on other neurons that they both project to. Neuronal synchrony has been proposed to underlie many processes related to perception and attention.

THE LOCALIZATION OF FUNCTION

the 30-second neuroscience

RELATED BRAINPOWER
See also
THE BASIC ARCHITECTURE
OF THE BRAIN
page 24

NEUROPSYCHOLOGY
page 56

BRAIN IMAGING
page 58

THE HUMAN CONNECTOME
page 60

3-SECOND BRAINWAVE
Cognitive functions are neither fully localized nor wholly distributed in the brain – each function depends on a complex but specific network of interacting brain regions.

3-MINUTE BRAINSTORM
The question is not *where* in the brain a cognitive function sits, but *what* are the underlying mechanisms and key interacting networks supporting it. Neuroscientists have discovered that, while there is clear functional specialization in the brain, these dedicated areas do not work alone – they are better thought of as hubs within complex interconnected networks. For example, fear in the brain depends on the amygdala. If this area is removed, a person becomes fearless, but it is the amygdala's connections and extended network that allow us to really feel fear.

Like cartographers of the mind, nineteenth-century scientists began to map the *terra incognita* of the brain. Franz Gall, founder of the theory of phrenology, proposed that different parts of the brain produced bumps in the skull depending on individual variation in mental capability and personality. He was proven wrong, but the questions underlying his ideas endured: do memory, language, attention, emotion and perception depend on particular brain areas or are these cognitive functions distributed across the brain? The main way of testing this idea was 'ablation', the destruction of a specific area of the brain in an animal (or analyzing people with specific brain lesions). While in humans this evidence seemed to support the localization of function, in animals the case was less clear. Karl Lashley trained rats to navigate around a maze and tested their behaviour after damaging specific parts of their brains. He found that the reduction in performance depended more on the amount of brain tissue damaged than on the specific location affected. This led him to defend the idea of 'mass action in the brain', which argued that the cerebral cortex acts 'as a whole' in many different types of learning. The modern consensus is that while many functions are indeed associated with particular brain areas, each function nonetheless depends on interactions in widely distributed networks involving many different areas.

3-SECOND BIOGRAPHIES
FRANZ GALL
1758–1828
Founder of phrenology

KARL LASHLEY
1890–1958
Defender of the ideas of 'mass action' and 'equipotentiality' in the brain

30-SECOND TEXT
Tristan Bekinschtein

An artist's impression of local networks in the brain – now you know how lab rats in a maze feel. There will be cheese if you make it through.

HEBBIAN LEARNING

the 30-second neuroscience

3-SECOND BRAINWAVE
What happens when
we learn something new?
'Neurons that fire together,
wire together.'

3-MINUTE BRAINSTORM
The space between two
neurons is called the synaptic
cleft. The neurotransmitters
that use this little gap as a
bridge may make the first
neuron excite or inhibit the
second. Learning works as
a combination of these two
types of communication.
Based on this, evidence of
learning has been found,
even in sea slugs. In this case,
the molecular mechanism
involves glutamate (a
neurotransmitter) released
by the first neuron, which
then crosses the gap and
binds to receptors in the
second neuron, which in
turn makes more receptors
available for the next time
the glutamate arrives.

What happens in the brain

when we learn something new? How do changes
in neurons and synapses lead to the formation
of new memories? Back in 1949, Donald Hebb
speculated that learning and memory might
depend on a simple process in which 'neurons
that fire together, wire together'. Think of this as
a trace, like footsteps left in snow, that deepens
whenever two neurons communicate. Early on,
Hebb's theory was applied to experiments on
conditioning that originated with Ivan Pavlov. For
example, a particular neuron in a bee's brain fires
when the bee is given sugar, causing the bee to
stick out her tongue (proboscis). If we introduce a
lemon odour before giving the bee the sugar, and
repeat this several times, the bee will start to stick
her tongue out when she smells the lemon, even
when no sugar is offered. This neuron now
fires to the odour alone: Hebbian learning has
strengthened the connections between these
neurons and others responding to lemony smells.
The power of Hebb's idea lies with evidence
showing that learning changes the connections
between two neurons at the molecular level. Now,
imagine you learn a new word: 'brain-numb'. You
are now creating new connections in the networks
of your linguistic brain through Hebbian learning.

RELATED BRAINPOWER
See also
NEURAL NETWORKS
page 40

THE NEURAL CODE
page 42

THE REMEMBERING BRAIN
page 118

3-SECOND BIOGRAPHIES
IVAN PAVLOV
1849–1936
Developed the concept of
conditioning, probably an
underlying mechanism of
learning in most of its forms

DONALD HEBB
1904–85
Proposed a mechanism for
learning in neurons

30-SECOND TEXT
Tristan Bekinschtein

*In a Hebbian-learning
experiment, bees stuck
out their tongues when
they smelled lemon,
because their neurons
had learned that sugar
will follow.*

NEURAL NETWORKS

the 30-second neuroscience

Imagine a computer that can
think like us. Now constrain its circuits to work
like networks of neurons giving birth to emergent
processes that underlie perception, thoughts
and actions. This is the aim of computational
neuroscience, and although it still remains out
of reach, the perspective of neural networks has
been highly influential in theories about how the
brain works. In 1890, William James proposed that
our thoughts are a product of interaction among
neurons, but this idea was very hard to test at
the time. Taking a more formal approach, in 1943
Warren McCulloch and Walter Pitts created a
mathematical model of a single neuron with
inputs and outputs that formed the basis of
the first artificial neural networks. In the 1970s,
mathematical arrays of artificial neurons were
created that started to mimic biological brain
mechanisms. These 'connectionist networks' had
new learning algorithms at their core and could
solve complex problems of pattern recognition.
Not only did these devices help scientists
understand how the brain might work, they also
led – and are still leading – to new biologically
inspired technologies. A key implication of this
view is that information is not represented locally
in the brain but is distributed across all the
connections in a network. The hunt is now on
to devise new neural network architectures that
mimic the brain even more closely.

3-SECOND BIOGRAPHIES
WARREN STURGIS MCCULLOCH
& WALTER PITTS
1898–1969 & 1923–69
Showed mathematically that
artificial neural networks can
implement logical, arithmetic
and symbolic functions

DAVID RUMELHART &
JAMES MCLELLAND
1942–2011 & 1948–
Created the first model with
formal rules that became the
basic framework for most
connectionism models

30-SECOND TEXT
Tristan Bekinschtein

*On the left, a neat
network of artificial
neurons; on the right,
a wrinkled sponge full
of water, sugar and
fat. Which is smarter?*

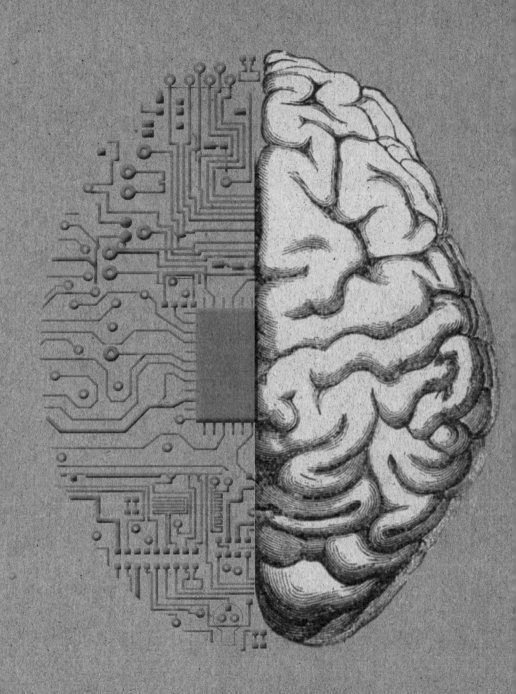

THE NEURAL CODE

the 30-second neuroscience

3-SECOND BRAINWAVE
How does the brain talk to itself? To unravel the mystery of the brain's operation, we will need to comprehend the language of neurons, individually and collectively.

3-MINUTE BRAINSTORM
One promising concept for neural coding is that of synchrony. Neurons that fire spikes together will probably have more impact on their targets than those that do not, assuming that the synchronously emitted spikes arrive at their destinations at the same time. Synchrony among large populations of neurons can be seen in the brain rhythms detectable using EEG and it has been suggested that these rhythms provide 'windows of opportunity' within which neurons can effectively communicate with each other.

How can a group of neurons

sense a change in the world, turn that into information, and pass it to other neural networks to generate perception and behaviour? Let us first be clear that the brain does not have one language but many, like a neuronal tower of Babel. Even worse, there may be different languages at different levels: neurons, ensembles (groups of neurons) and the whole brain. For example, simultaneously reading electrical activity from a cluster of neurons in a monkey's motor cortex shows that the simple sum of the different neurons' voices does not clearly explain the trajectory of the monkey's arm movement. But looking at the overall activity of the entire population of neurons does encode the trajectory, demonstrating that the integrated activity of tens of thousands of neurons is needed for a simple movement. How do distant groups of neurons coordinate their activity? Is the neural code a 'rate code', in other words, is it based on the speed at which neurons fire spikes? Or does the brain use a 'timing code', in which it is the precise firing pattern that matters? This is an old debate and the story seems to be more complex than 'one or the other'. The modern neural code breakers are using new mathematical tools and cutting-edge theories to show how rate and timing codes might work together in mediating the electrical conversations of the brain.

RELATED BRAINPOWER
See also
NEURONS & GLIAL CELLS
page 16

NEURAL NETWORKS
page 40

THE OSCILLATING BRAIN
page 46

3-SECOND BIOGRAPHIES
ALAN TURING
1912–54
Father of computer science, cracked the Enigma machine code

WILLIAM BIALEK
1960–
Formalized the problem of reading the neural code using information theory

30-SECOND TEXT
Tristan Bekinschtein

Scenes from Brain Central: 'Can you hear me? Please hold the line. Your call is important to us.'

22 July 1904
Born in Chester,
Nova Scotia, Canada

1925
Graduated with a BA
from Dalhousie
University

1932
MA in psychology
from McGill University;
his thesis entitled
'Conditioned and
Unconditioned Reflexes
and Inhibition'

1933–34
Wrote 'Scientific
Method in Psychology:
A theory of epistemology
based in Objective
Psychology'; unpublished,
but a mine of ideas

1934
Worked under Karl
Lashley at University
of Chicago

1935
Followed Lashley to
Harvard to continue
his doctorate on the
effects of early visual
deprivation on
perception in rats

1936
Awarded PhD

1937
Fellowship at the
Montreal Neurological
Institute under Wilder
Penfield

1942
Began working at Yerkes
Laboratory, studying
emotional processes
in the chimpanzee

1947
Appointed Professor of
Psychology at McGill
University, where he
remained until he
retired in 1972

1949
*The Organization
of Behavior: a
Neuropsychological
Theory* was published

1960
Appointed President
of the American
Psychological
Association

1972
Retired from McGill
University, but remained
as Emeritus Professor
of Psychology

1980
Returned to Dalhousie
University as Emeritus
Professor of Psychology

20 August 1985
Died in Nova Scotia

DONALD HEBB

Considering the impact he subsequently had on his chosen discipline, Donald Olding Hebb came somewhat late to psychology. His first ambition was to be a writer. Born in Nova Scotia, Canada, he began his academic life at Dalhousie University (after an indifferent school career), where he achieved a BA. His writing career did not take off, so he made the pragmatic move into education, teaching for some years in elementary and high schools. His reading of Freud and Pavlov engendered an interest in psychology, and he studied for an MA at McGill University, followed by a PhD from Harvard (under Karl Lashley), a stint at the Montreal Neurological Institute with Wilder Penfield and some time at Yerkes Laboratory of Primate Biology.

Hebb's work spanned neurophysiology and psychology. His overriding interest was in the link between brain and mind – how neurons behave and arrange themselves to produce what we perceive as thoughts, feelings, memories, and emotions. He came at the conundrum from all directions. He studied sensory deprivation, brain damage, the effects of brain surgery, behaviour, experience, environment, stimulation and heredity, as well as the major theories in psychology of the time, including Gestalt and behaviourism, and the work of Freud, Skinner and Pavlov. His findings inspired him to formulate a kind of grand unified theory for neuroscience that he hoped would unite brain and mind.

In 1942, while at Yerkes, he began writing his seminal work, *The Organization of Behavior: A Neuropsychological Theory*, setting out his ideas. It introduced the concept of the Hebbian synapse (broadly, the idea that 'neurons that fire together, wire together') and the cell assembly, the notion that Hebbian learning would lead to groups of neurons activating in particular sequences, enabling thought, perception, learning and memory. The work was published in 1949, a year after he was appointed Chair of Psychology at McGill University. Its influence and relevance was far-reaching and Hebb's pioneering work remains the basis for many developments in robotics, computer science, artificial intelligence and engineering as well as neuroscience and developmental psychology.

Both neuroscientists and psychologists claimed Donald Hebb as one of their own. It is a tribute, maybe, to his latent writing skills that he was able to look forensically at the detail of his research and at the same time step back and show his readers the bigger picture – like all great novelists, a unifier of the general and the particular.

THE OSCILLATING BRAIN

the 30-second neuroscience

3-SECOND BRAINWAVE
Synchrony among large populations of neurons leads to characteristic brain oscillations that may underlie many perceptual, cognitive and motor functions.

3-MINUTE BRAINSTORM
Brainwaves are not always good for you. When neural oscillations get too strong, the brain can enter a state of so-called 'hypersynchrony', where vast swathes of neural real estate turn on and off together. This kind of electrical storm – literally a 'brainstorm' – is what happens during epileptic seizures. Being able to predict hypersynchronous episodes before they happen, perhaps in time to allow an intervention, is a major current challenge for clinical neuroscientists.

Just how often do you have a
brainwave? If you are having a bad day you may think 'hardly ever', but the truth is that you are having brainwaves all the time, even while you are asleep. The electrical activity of the brain, when measured by techniques such as EEG, is characterized by strong oscillations – waves – which are thought to arise from synchronous activity in large populations of neurons. In fact, 'alpha waves' were about the first thing to be noticed by Hans Berger when he invented the EEG back in the 1920s. Alpha waves are relatively slow oscillations at about 10 Hz (10 cycles per second), which are observed predominantly across the back of the brain and are most pronounced during relaxed wakefulness with the eyes closed. This has led some researchers to suggest that the alpha rhythm reflects cortical 'idling', though this view is now being challenged. Other prominent brain oscillations include the delta (1–4 Hz), theta (4–8 Hz), beta (12–25 Hz) and gamma (25–70 Hz or higher) rhythms, however, these categories are a little arbitrary. The search is now on to discover what these different oscillations do for the brain and mind. For example, beta oscillations appear over cortical motor areas as the brain prepares for movement, and the gamma rhythm has long been argued to support the binding of perceptual features together in complex scenes.

RELATED BRAINPOWER
See also
BRAIN IMAGING
page 58

SLEEP & DREAMING
page 78

3-SECOND BIOGRAPHIES
HANS BERGER
1873–1941
Neurophysiologist who was the first to record EEG signals; alpha waves were originally called 'Berger waves'

WILLIAM GREY WALTER
1910–77
Neurologist and cybernetician who was the first to measure delta waves during sleep

WOLF SINGER
1943–
Pioneered the idea that gamma oscillations may play key roles in perception

30-SECOND TEXT
Anil Seth

Hans Berger surfs his own brainwaves. He can assure you that it is a lot more invigorating than it may look.

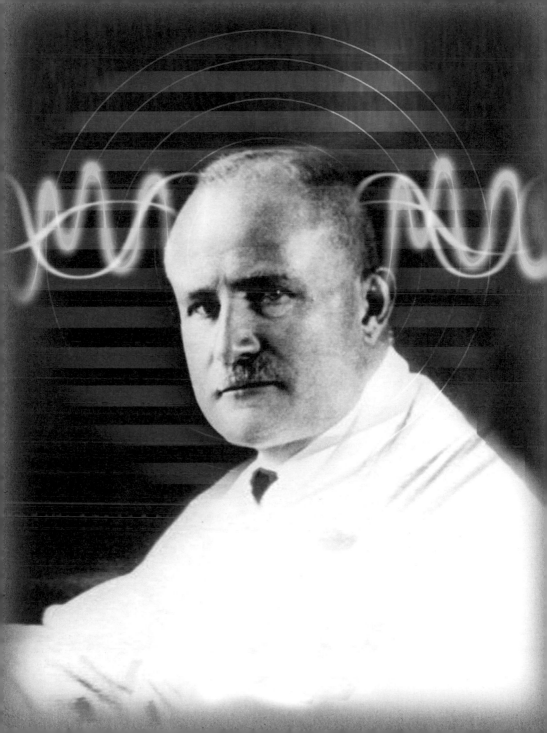

NEURAL DARWINISM

the 30-second neuroscience

3-SECOND BRAINWAVE
Population thinking, so
fundamental for Darwin,
may also be the key to
understanding the brain.

3-MINUTE BRAINSTORM
Neural Darwinism
draws a key distinction
between 'instruction' and
'selection'. Instructionist
systems, like standard
computers, rely on programs
and algorithms, and they
suppress variability and
noise. Selectionist systems
depend on large amounts of
variation and involve the
selection of specific states
from very large repertoires.
So the theory provides a
sharp contrast to computer
models of brain and mind,
highlighting – as Darwin
had beforehand – that
variation is essential to
biological function.

Darwin's theory of natural
selection is one of science's greatest
accomplishments, explaining how complex life
emerges from evolutionary variation and selection
over very long time spans. Neural Darwinism,
developed by Gerald Edelman, proposes that a
similar process may occur in the brain, involving
groups of neurons instead of genes or organisms.
Also known as the theory of neuronal group
selection (TNGS), it rests on three proposals.
The first is that early brain development generates
a highly diverse population of neuronal circuits.
The second is that there is selection among
these groups: those that are used survive and
strengthen; those that aren't wither away. Finally,
the TNGS proposes the idea of 're-entry' – a
constant interchange of signals between widely
separated neuronal populations. Francis Crick
criticized the theory, pointing out that it lacked
a mechanism for replication, a key property
of natural selection alongside diversity and
selection. However, Edelman has continued to
develop his theory, providing new accounts of
language and even consciousness. Although
direct evidence is still lacking, TNGS seems again
increasingly relevant as neuroscientists struggle
to understand how very large populations of
neurons behave and develop.

RELATED BRAINPOWER
See also
THE DEVELOPING BRAIN
page 28

HEBBIAN LEARNING
page 38

NEURAL NETWORKS
page 40

3-SECOND BIOGRAPHIES
GERALD M. EDELMAN
1929–2014
Nobel Laureate in Physiology
or Medicine for his work on
selectionist principles in
immunology, has been actively
developing neural Darwinism
since the 1970s

JEAN-PIERRE CHANGEAUX
1936–
Also a pioneer in developing
selectionist theories of brain
function; in other seminal work
he discovered and described how
nicotine acts within the brain

30-SECOND TEXT
Anil Seth

*Brain pruning – some
neurons can fly, others
go the way of the dodo.*

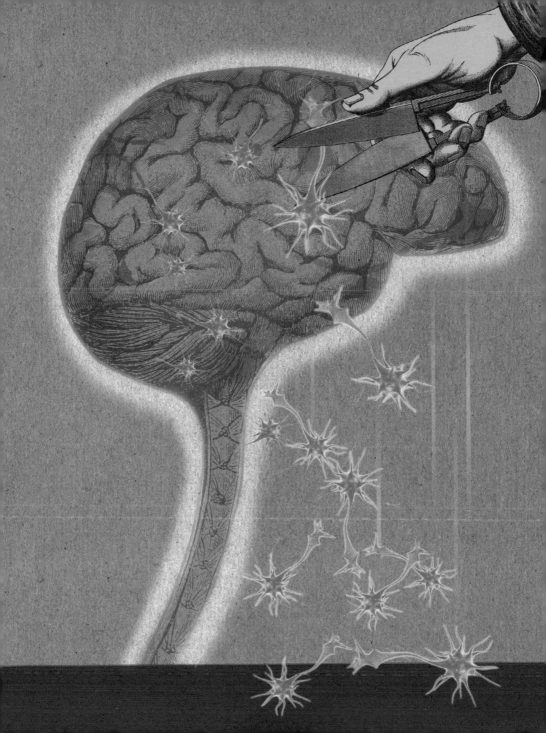

THE BAYESIAN BRAIN

the 30-second neuroscience

3-SECOND BRAINWAVE
Perceiving is believing.
The perceived world is the
brain's best guess of the
causes of its sensory inputs.

3-MINUTE BRAINSTORM
The Bayesian brain theory
implies surprisingly deep
connections between
perception and imagination.
The theory requires that
the brain maintains a
'generative model' of the
causes of its sensory inputs.
Put simply, this means
that in order to perceive
something, the brain must
be able to self-generate
corresponding perception-
like states from the 'top
down'. If this is true,
it means each of our
perceptual worlds depends
on our own individual
imaginative abilities.

Imagine being a brain. You are
stuck inside a bony skull, trying to figure out
what's out there in the world. All you have to
go on are streams of electrical impulses from the
senses, which vary depending on the structure
of that world and, indirectly, on your own outputs
to the body (move your eyes and sensory inputs
change, too). In the nineteenth century, Hermann
von Helmholtz realized that perception – the
solution to the 'what's out there' problem – must
involve the brain inferring the external causes of
its sensory signals. This suggests that brains
perform something like 'Bayesian inference',
a term that describes how beliefs are updated as
new evidence comes in. In other words, incoming
sensory data are combined with 'prior beliefs'
to determine their most probable causes, which
correspond to perceptions. At the same time,
differences between predicted signals and actual
sensory data – 'prediction errors'– are used to
update the prior beliefs, ready for the next round of
sensory inputs. One interpretation of this idea –
predictive coding – argues that the architecture of
the cortex is ideally suited for Bayesian perception.
On this view, 'bottom-up' information flowing
from the sensory areas carries prediction errors,
while 'top-down' signals from higher brain
regions, convey predictions.

RELATED BRAINPOWER
THE BASIC ARCHITECTURE
OF THE BRAIN
page 24

THE IMAGINING BRAIN
page 122

THE SCHIZOPHRENIC BRAIN
page 150

3-SECOND BIOGRAPHIES
THOMAS BAYES
1701–61
British theologian and
philosopher credited with
formulating the basic logic
of what is now known as
Bayes' theorem

HERMANN VON HELMHOLTZ
1821–94
German physiologist and
physicist who formulated
the principle of 'perception as
inference', and was also the
first to measure the speed of
electrical impulses in nerves

30-SECOND TEXT
Anil Seth

*Less 'what you see
is what you get',
more 'what you get
is what you see'* .

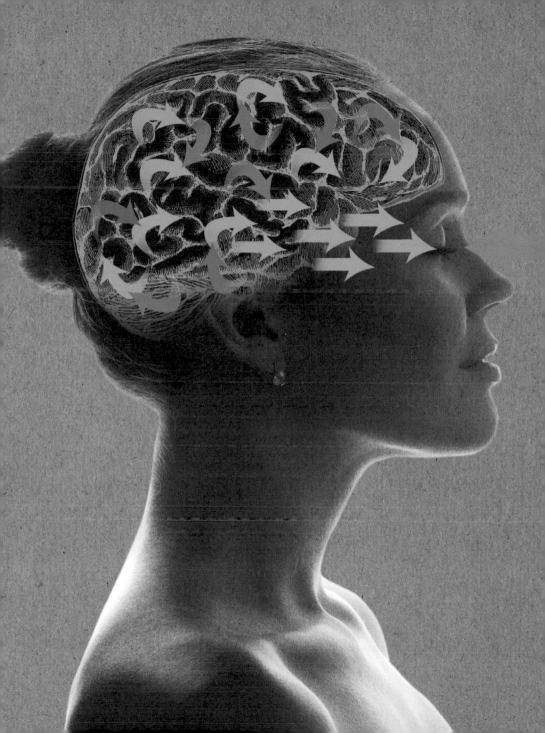

MAPPING THE BRAIN

connectome A term coined by Olaf Sporns, by analogy with the genome (the map of genes), the connectome is the map – or wiring diagram – of all the connections in the brain. While the broad outlines of the human connectome are known, we are very far from unravelling the connectome in all its detail.

cerebral cortex The deeply folded outer layers of the brain, which take up about two-thirds of its entire volume and are divided into left and right hemispheres that house the majority of the 'grey matter' (so- called because of the lack of myelination which makes other parts of the brain seem white). The cerebral cortex is separated into lobes, each having different functions, including perception, thought, language, action and other 'higher' cognitive processes such as decision making.

default mode network (DMN) A group of brain regions whose activity (typically when measured by fMRI) is reduced during the performance of an externally directed task, and is more active in states of wakeful rest, mind-wandering, introspection or inwardly directed attention. In general, the DMN has been associated with self-related processing. It includes medial parts of the prefrontal and temporal lobes and the posterior cingulate cortex.

diffusion tensor imaging (DTI) DTI is a relatively recent neuroimaging technique that uses magnetic resonance imaging (MRI) to chart the long-range bundles of connections (axons) that course throughout the brain. The method depends on the fact that water molecules diffuse preferentially along axons, instead of across them.

electroencephalography (EEG) The practice of detecting the tiny variations in electrical field at the surface of the brain, which are produced by the activity of populations of neurons in the underlying cortex. EEG has very good resolution in time but is relatively poor (as compared to fMRI) in localizing activity in space. A related method – magnetoencephalography (MEG) – measures the corresponding magnetic field variations. MEG can be more sensitive than EEG but is a much more complex and expensive technology.

frontal lobes One of the four main divisions of the cerebral cortex and the most highly developed in humans as compared with other animals. The frontal lobes (one for each hemisphere) house areas associated with decision making, planning, memory, voluntary action and personality.

(functional) magnetic resonance imaging (f)MRI MRI technology has revolutionized neuroscience by allowing non-invasive mapping of the three-dimensional structure of the brain, taking advantage of the way different parts of the brain react under strong magnetic fields. fMRI extends MRI to measure brain activity and is based on measuring the differences in blood oxygenation that go along with neural activity. fMRI has very good spatial resolution but poor time resolution, as compared to EEG.

neurons The cellular building blocks of the brain. Neurons carry out the brain's basic operations, taking inputs from other neurons via dendrites, and – depending on the pattern or strength of these inputs – either releasing or not releasing a nerve impulse as an output. Neurons come in different varieties but (almost) all have dendrites, a cell body (soma) and a single axon.

neuropsychology This is the discipline of inferring the function of different brain regions based on the behaviour and reported experiences of patients who have experienced damage to specific regions. For example, the amnesia in patient H.M. following hippocampal damage allowed neuropsychologists to associate the hippocampus with (episodic) memory.

phrenology Popularized by Franz Gall in the nineteenth century, phrenology is the now discredited practice of inferring personality and mental attributes from the various lumps and bumps on the surface of the skull. Although he was wrong about this, Gall was very much right in the idea that different parts of the brain did different things, thus laying the foundations for neuropsychology and even modern fMRI.

transcranial magnetic stimulation (TMS) A technique in which short but powerful magnetic pulses are applied to the scalp, briefly stimulating the neurons in the underlying cortex. By perturbing brain activity in specific regions and observing what happens, TMS can help determine the function of these regions. Recently, TMS has been combined with EEG so that brain as well as behavioural responses to TMS pulses can be recorded.

NEUROPSYCHOLOGY
the 30-second neuroscience

3-SECOND BRAINWAVE
Science can link function to brain region, via patients unfortunate enough to experience brain damage along with a discrete mental impairment.

3-MINUTE BRAINSTORM
Neuropsychology was always a messy business. For instance, brain damage rarely creates a neat, discrete lesion, while intact regions can sometimes take over a given function from a damaged one. But for many decades, this method was the main brain-mapping tool in town. Now, though, neuropsychology has fallen somewhat out of fashion, with modern brain-scanning techniques, such as fMRI efficiently able to search across the whole brain for links to a specific function in healthy subjects.

Brain mapping started shakily in the late eighteenth century with the pseudoscience of phrenology, which attributed bumps on the skull to specific psychological traits. Partly to refute phrenology, Paul Broca published a landmark study in 1861, in which he reported on a patient, Leborgne, who understood what was said to him but whose speech had so degraded that he could only ever say one word, 'tan'. Leborgne had just died and the autopsy carried out by Broca found local damage in a restricted part of the left frontal lobe. Although this was only one patient, it was, crucially, direct evidence to link function with region. Broca went on to find many speech-impaired patients with the same damaged brain area. Later, Carl Wernicke added to this picture by using the same method to link understanding of language with a portion of the left temporal lobes. These cases helped give birth to a new kind of science, that of neuropsychology, in which brain-damaged patients are examined for deficits, enabling us to learn what regions are crucial for a given function. Over the past 150 years, countless brain-damaged patients have helped build up a picture of a brain with many specialist subunits, each playing their part in our thoughts and feelings.

RELATED BRAINPOWER
See also
THE LOCALIZATION
OF FUNCTION
page 36

RRAIN IMAGING
page 58

BRAIN STIMULATION
page 70

3-SECOND BIOGRAPHIES
PAUL BROCA
1824–80
Discovered the speech production area

CARL WERNICKE
1848–1905
Discovered the speech comprehension area

ALEXANDER LURIA
1902–77
Father of modern neuropsychology

30-SECOND TEXT
Daniel Bor

Phineas Gage and the tamping iron that went right through his brain, leaving him alive but changing his mind.

BRAIN IMAGING

the 30-second neuroscience

3-SECOND BRAINWAVE
Brain imaging allows us to study the shape, wiring and function of the human brain with great precision in a safe, non-invasive way.

3-MINUTE BRAINSTORM
fMRI is starting to be used as a mind-reading device. So far, this is largely limited to estimating which of a small set of pictures, video snippets or words was just presented to an individual. But new methods are being developed that actually reconstruct perception by reading activity in the visual cortex and generating a fuzzy image from this. It is tantalizing to contemplate how far this form of technological telepathy will progress in the future.

Just as astronomy and biology were revolutionized by the telescope and microscope, neuroscience has been transformed by brain-imaging technologies. Some scan types, such as magnetic resonance imaging (MRI), non-invasively reveal the three-dimensional structure of the brain. These are useful for exploring how the brain is constructed and how neural anatomy differs among people. They also provide vital clinical tools for detecting various kinds of brain damage. A more recent development is diffusion tensor imaging (DTI), which provides a three-dimensional map of the main wires connecting brain regions together. But what has truly revolutionized neuroscience are technologies that observe the brain's activity. Electroencephalography (EEG) has played a part in this by revealing the brain's changing patterns of electrical activity, millisecond by millisecond, as we perform tasks or undergo different sleep stages. However, EEG is poor at attributing functions to specific brain regions. By far the most dominant imaging technology over the last two decades has been functional magnetic resonance imaging (fMRI), which can pinpoint neuronal activity changes to within a few millimetres and a few seconds. This is dramatically improving our understanding of the functional role of each brain region and how areas collaborate to support mental processes.

RELATED BRAINPOWER
See also
THE LOCALIZATION
OF FUNCTION
page 36

NEUROPSYCHOLOGY
page 56

RESTING STATE
page 66

3-SECOND BIOGRAPHIES
HANS BERGER
1873–1941
Pioneer of EEG

PAUL LAUTERBUR
1929–2007
Pioneer of MRI

PETER MANSFIELD
1933–2017
Pioneer of MRI

30-SECOND TEXT
Daniel Bor

Other people can read your mind. Confirm your privacy settings by thinking only of primary colours and simple geometric shapes.

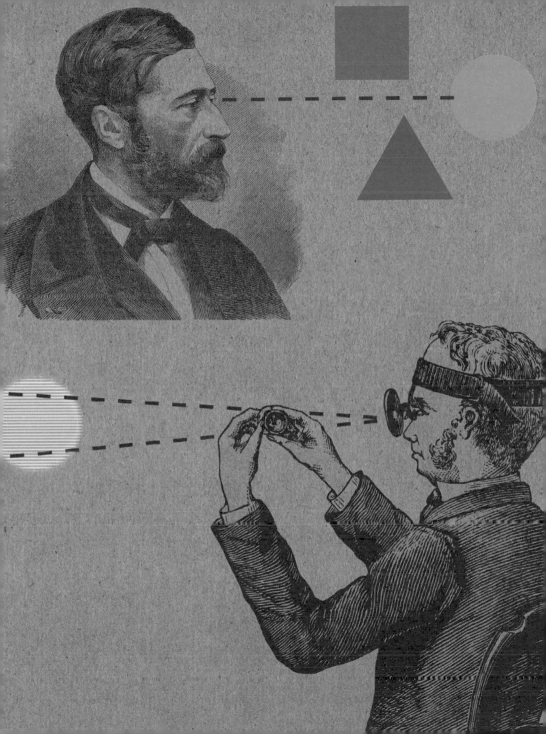

THE HUMAN CONNECTOME

the 30-second neuroscience

3-SECOND BRAINWAVE
The human connectome is the entire map of all the 600 trillion wires in a human brain. We remain decades away from completing this project.

3 MINUTE BRAINSTORM
Some believe that revealing the network structure of the human brain might yield fewer clues than promised, especially given that any connectome is continuously in flux as wires grow or die. One species in which the connectome has been largely complete for many years is that of the nematode worm, *Caenorhabditis elegans*. With a brain of just 302 neurons, it is one of the simplest animals, and yet many of its behavioural features remain a mystery.

At whatever level you look, from the microscopic wires of a handful of neurons, up to the finger-thick fibres that connect major regions of cortex, the brain is essentially structured as a network. The entire map of all these networked wires is known as the connectome. Various ambitious, large-scale projects around the world are starting to piece together the human connectome from different angles. At the cellular level, much of this involves painstaking microscope work on tiny anatomical sections. Larger scale, though far cruder, methods include a form of MRI scanning technology called DTI, which is designed to create an image of the brain's major pathways non-invasively. One particular challenge in this global project is to knit together the various different techniques to create a coherent overall picture of the brain's wiring. Although the activity of our many billions of neurons, along with our brain chemistry and genetics, are essential shapers of our mental world, many neuroscientists now think that this network structure is the most critical feature of all. The hope is that mapping the human connectome and exploring how it differs between people will reveal vital clues about our thoughts, the nature of psychiatric illnesses and, ultimately, who we are as mental beings.

RELATED BRAINPOWER
See also
NEURAL NETWORKS
page 40

BRAIN IMAGING
page 58

3-SECOND BIOGRAPHIES
DAVID VAN ESSEN
1945–
Leader of the Human Connectome Project and pioneer in neuroanatomy

CORNELIA BARGMANN
1961–
American neurobiologist known for her work on *C. elegans*

OLAF SPORNS
1963–
German scientist who coined the term 'connectome'

30-SECOND TEXT
Daniel Bor

High-tech in the head. A two-dimensional rendition of a three-dimensional DTI image of some of the 600 trillion wires that power your brain.

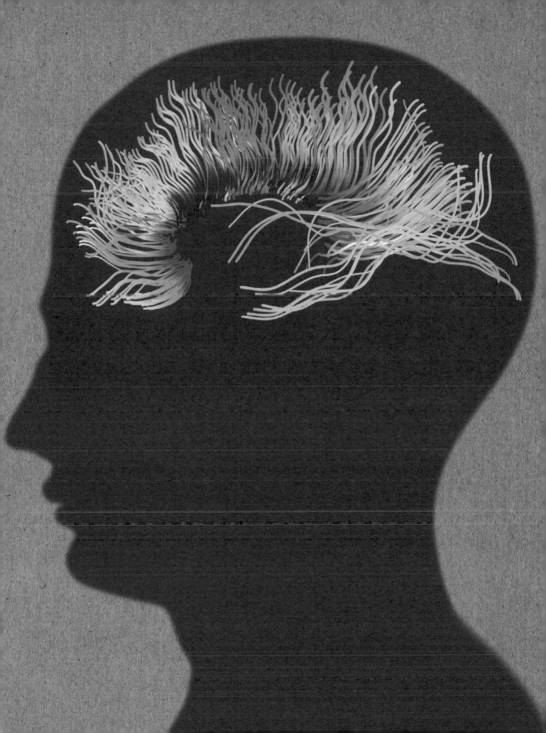

OPTOGENETICS

the 30-second neuroscience

3-SECOND BRAINWAVE
Optogenetics involves genetically altering neurons, so that they can be precisely manipulated by using light to turn them on and off at will.

3-MINUTE BRAINSTORM
What if you could immediately calm a raging storm of overactive epileptic neurons, just with a powerful lamp? Or take dormant movement-controlling cells that leave patients with Parkinson's disease fighting just to make the simplest of movements and reinvigorate these neurons with an implanted micro-flashlight? Although significant safety concerns have to be met before human trials start, the potential applications for a wide range of psychiatric and neurological conditions are breathtaking.

In 1999 Francis Crick, the co-discoverer of DNA, mentioned some 'far-fetched' ideas about genetically engineering neurons so that light alone could switch them on and off. He should have had more faith in his own field. Within five years, Karl Deisseroth and colleagues had used a virus to smuggle an algal light-sensitive gene into rat neurons. When exposed to a blue light, the neurons fired. Soon an 'off' as well as an 'on' switch was found. Another gene, this time from bacteria, could be added in a similar way, so neuronal firing could be suppressed whenever green light was shone on it. Now hundreds of labs around the world use similar techniques to probe the machinery of the brain with unprecedented control. Sometimes light can be shone from the surface, on the skull, but usually tiny emitters are implanted deep inside the brain. This technique can even be used to boost cognitive performance – Wim Vanduffel and colleagues recently modified a region of the frontal lobes in two monkeys to enable widespread light-induced activation. When a blue light turned this region on, the monkeys were faster at an object-tracking task. Therefore, as well as being one of the most innovative neuroscientific techniques of recent times, optogenetics holds great promise as a future clinical tool.

RELATED BRAINPOWER
See also
THE LOCALIZATION OF FUNCTION
page 36

NEUROPSYCHOLOGY
page 56

BRAIN IMAGING
page 58

BRAIN STIMULATION
page 70

3-SECOND BIOGRAPHIES
FRANCIS CRICK
1916–2004
First scientist to suggest optogenetics as a technique

KARL DEISSEROTH
1971–
Leading pioneer of the technique of optogenetics

ED BOYDEN
1979–
Collaborator with Deisseroth

30-SECOND TEXT
DANIEL BOR

You know that light-bulb moment? Maybe it isn't just a metaphor.

26 January 1891
Born Spokane,
Washington

1899
Moved with his family to
Hudson, Wisconsin, his
mother's home

1913
Graduated from
Princeton

1914–16
Rhodes Scholar at
Merton College Oxford;
studied neuropathology
under Charles
Sherrington

1918
MD from Johns Hopkins
University; served
apprenticeship under
brain surgeon Harvey
Cushing in Boston,
Massachusetts

1919
Final year in Oxford as a
Rhodes Scholar, followed
by study in Europe

1921
Returned to USA to
become Associate
Surgeon at Columbia
University

1928
Joined the Medical
Faculty at McGill
University

1934
Acquired funding for,
founded and became
director of the Montreal
Neurological Institute
(MNI), part of McGill
University

1934
Became a Canadian
citizen

1951
Wrote, with Herbert
Jasper, *Epilepsy and the
Functional Anatomy
of the Human Brain*

1954
Retired from McGill
Medical Faculty but
continued as Director
of the MNI

1960
Awarded the Lister
Medal for contributions
to surgical science

5 April 1976
Died in Montreal,
Canada

WILDER PENFIELD

A pioneering brain surgeon and probably neuroscience's greatest team player, Wilder Penfield was born in the USA and raised in Hudson, Wisconsin, but he claimed Canadian citizenship via his mother in 1934. He was a stalwart football star at Princeton (part of a career plan to gain a Rhodes Scholarship, which depended on a manly mixture of sporting and intellectual prowess) and spent a year after graduation as the team coach to help finance further study. Hard work and training paid off and in 1914 Penfield was awarded a Rhodes Scholarship to Merton College Oxford in England. He studied under neurophysiologist Charles Sherrington, who opened his mind to the uncluttered pastures of neuroscience, where there was much to be explored.

On his return to the USA, Penfield embarked on his career as a neurosurgeon, reasoning that he could better carry out research into the functions and secrets of the human brain if he had one under his scalpel. Penfield had a strong team ethic and philanthropic drive, apparently instilled by his mother. Instead of working alone, he envisaged an entire institute in which neuroscientists of all disciplines could work, research and learn together and share their findings for the betterment of humanity. The thinking at the time in New York did not suit this model, so he moved to the Medical Faculty of McGill University, from which base he lobbied energetically for funds, securing a grant from the Rockefeller Foundation. In 1934, he set up the Montreal Neurological Institute, which would become a powerhouse of neuroscientific research.

It was here that Penfield did the work on epileptic patients for which he is best remembered. He introduced the Montreal Method, in which he operated on patients to excise the parts of their brain from which epileptic seizures originated. He did this under local anaesthetic so that, invaluably, they could respond to his questions as he operated. His patients reported that when different parts of the brain were probed, they experienced different feelings and sensations. From the first-hand information gained, Penfield was able to make preliminary maps of the brain, establish the principle of brain lateralization and lay the foundations for future brain mapping. The stylized 'homunculus' that he produced with his colleague Herbert Jasper, in which the size of each body part reflects the number of nerves that serve it, is still in use today.

RESTING STATE

the 30-second neuroscience

3-SECOND BRAINWAVE
Without a task to perform, the brain is almost as active as normal, but a 'default' set of regions lights up instead of task-related ones.

3-MINUTE BRAINSTORM
The default mode network is delicately defined by the coordinated activity of a range of cortical areas, mainly located along the 'cortical midline' – where the hemispheres of the brain touch in the middle. A proper default mode takes years to emerge in infancy, and old age commonly disrupts it. Further evidence of the importance of the default mode network comes from its most promising putative role, that of daydreaming, which has been linked both to higher insight and creativity.

In functional brain-imaging experiments, it is standard practice to give a volunteer a taxing task to perform while in the scanner and to associate the demands of the task with those parts of their brain that are observed to light up. But what happens in between those periods of effort? One might expect the brain's activity to dramatically drop and some random firing pattern to ensue. The first clue that such assumptions are wrong came in the early 2000s from Marcus Raichle, who found that there is a consistent group of brain regions (known as the default mode network) whose activity is suppressed when we perform any focused task, but which springs back into action when we can mentally 'twiddle our thumbs'. Around the same time, Michael Greicius uncovered the same set of regions when he took the bold step of deliberately scanning subjects when they were just resting. This striking clue about the brain's activity is in a sense the 'dark energy' of neuroscience and researchers are still searching for an adequate account of it. One explanation is that it may be the neural signature of daydreams. Intriguingly, abnormalities in the default mode network are associated with a range of disorders, so perhaps this seemingly trivial pastime is far more important than we originally thought.

RELATED BRAINPOWER
See also
THE LOCALIZATION
OF FUNCTION
page 36

BRAIN IMAGING
page 58

THE AGEING BRAIN
page 144

3-SECOND BIOGRAPHIES
MARCUS RAICHLE
1937–
Discoverer of the default mode network

MICHAEL GREICIUS
1969–
Carried out pioneering studies of resting state

30-SECOND TEXT
Daniel Bor

Slackers rejoice! The brain is working almost as hard when you are staring into space as it does when you are looking busy.

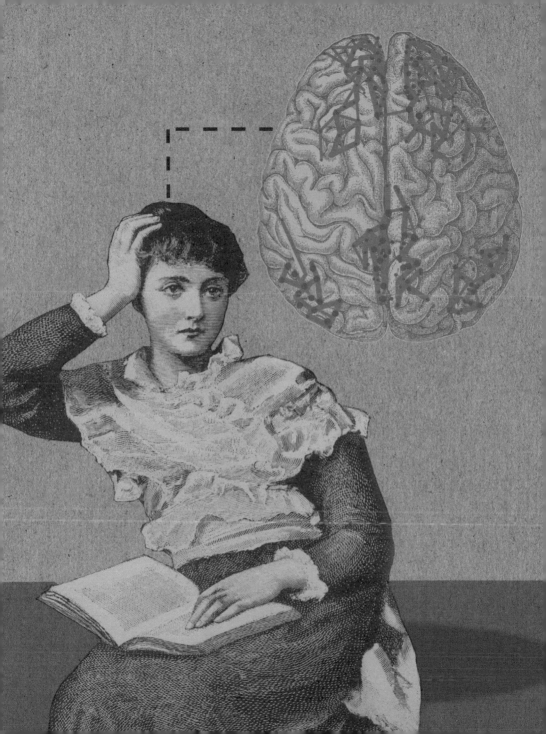

LEFT BRAIN VS RIGHT BRAIN

the 30-second neuroscience

3-SECOND BRAINWAVE
The two halves of the brain do work differently, but the notion of a creative right brain and logical left is an oversimplification.

3-MINUTE BRAINSTORM
An industry of apps and self-help books has grown up around the idea of unlocking the right brain's creative potential. There is evidence for right hemisphere creativity, but the left hemisphere is creative in its own way, too. Work with split-brain patients revealed the 'interpreter phenomenon' – the way the left hemisphere was very good at telling stories to explain what the left hand (controlled by the right hemisphere) was up to.

Looking at a human brain,
among the most obvious features is the fault line that runs along its centre from front to back, dividing the outer cortex into two distinct hemispheres. Although broadly anatomically symmetrical, the two halves don't function in the same way. Nineteenth-century physicians, such as Paul Broca, realized this because patients with damage to the left side were far more likely to have language problems than those with damage to the right. Interest in the issue was re-ignited in the 1960s when Roger Sperry and others began investigating 'split-brain' patients, who'd had the thick bundle of nerves connecting their hemispheres cut to treat severe epilepsy. Testing these patients showed that the two hemispheres could operate independently and had different strengths and weaknesses. Today, it is popular to characterize the left hemisphere as cold and logical and the right as emotional and creative. This is an over-simplification. Split-brain patients aside, most people's brain hemispheres work together. Instead of tasks being delegated to one side or the other, both hemispheres typically apply a different processing style to the same tasks. While the left hemisphere is dominant for language, the right has language functions of its own, including recognizing intonation.

RELATED BRAINPOWER
See also
THE LOCALIZATION OF FUNCTION
page 36

NEUROPSYCHOLOGY
PAGE 56

THE LINGUISTIC BRAIN
page 126

PAUL BROCA
page 124

ROGER SPERRY
page 148

3-SECOND BIOGRAPHIES
ROGER SPERRY
1913–94
Won the Nobel Prize for his work with split-brain patients

MICHAEL GAZZANIGA
1939–
Trained with Sperry to pioneer split-brain experiments

30-SECOND TEXT
Christian Jarrett

Geeks to the left of me, creatives to the right? Where does that leave Leonardo?

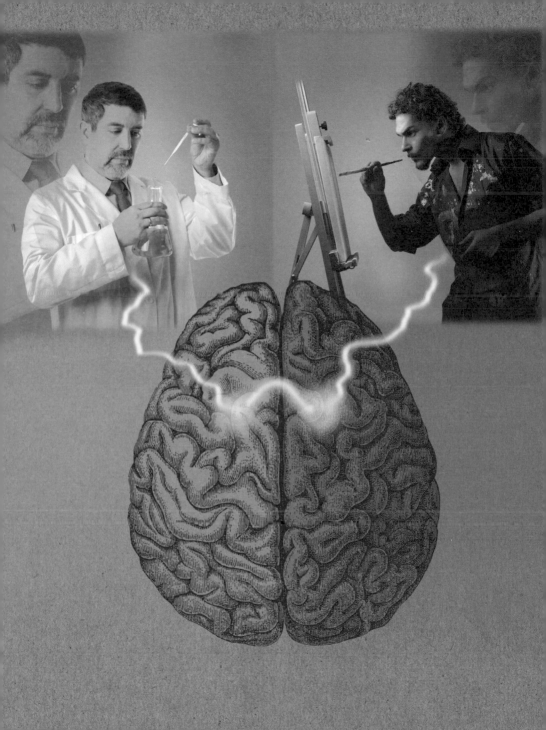

BRAIN STIMULATION

the 30-second neuroscience

3-SECOND BRAINWAVE
Neurons in specific locations can be electromagnetically stimulated, inducing changes in thought, perception or behaviour.

3-MINUTE BRAINSTORM
Wilder Penfield also found that stimulating one neuron might make the patient think his right cheek had been touched; while stimulating another might make his left thumb twitch. By repeated stimulations, he discovered that the motor and sensory cortices form a very ordered map, for instance with tongue movements controlled on the lower outer cortical section. This arrangement, broadly the same for us all, is one of the most pronounced cortical examples of the localization of function.

Wilder Penfield, one of the most influential neurosurgeons of the twentieth century, would commonly operate on severely epileptic patients while they were conscious (operating on the brain itself causes no pain). In order to minimize the amount of brain tissue requiring excision, he pioneered the use of an electrical probe to stimulate neurons and determine more precisely if they were part of the main abnormality causing seizures. He soon discovered that this method was also useful for mapping the functions of different brain regions and reported, for instance, that stimulation of a single neuron in the temporal lobes would reactivate entire memories in the patient. Transcranial magnetic stimulation (TMS) is a popular modern successor to this technique. TMS is a non-invasive procedure that uses a brief magnetic pulse on the scalp to stimulate the underlying cortical region (roughly a square inch or so in size). Depending on the technique, this can raise, or more usually suppress, the region's activity. If volunteers become better or worse at a particular task following TMS to a specific brain region, then this shows that the region is related to the corresponding process. In this way, TMS has become a useful additional tool to functionally map the brain.

RELATED BRAINPOWER
See also
THE LOCALIZATION
OF FUNCTION
page 36

NEUROPSYCHOLOGY
page 56

BRAIN IMAGING
page 58

3-SECOND BIOGRAPHIES
WILDER PENFIELD
1891–1976
Pioneered brain stimulation and discovered detailed neural sensory and motor maps

ANTHONY BARKER
1950–
First to use TMS in scientific research

JOHN ROTHWELL
1954–
Inventor of modern TMS technique to extend its effects by many minutes

30-SECOND TEXT
Daniel Bor

All this brain stimulation can make a guy a little twitchy.

CONSCIOUSNESS

autoscopic experience An autoscopic experience is the experience of seeing one's body from an external perspective. It is related to, but distinct from, out-of-body experiences, which involve a change in the perceived location of the self. Heautoscopy, an intermediate form, involves autoscopy with some uncertainty or alternation regarding perceived self-location.

binocular rivalry An experimental method in which each eye is presented with a different (incompatible) image, so that conscious perception alternates repeatedly between each. Because sensory input remains constant, examining neural activity during binocular rivalry can help reveal the brain basis of consciousness.

coma A major disorder of consciousness and the nearest thing to brain death before actually dying. Patients in a comatose state show no signs of wakefulness or awareness. Brain activity during coma is much reduced and in some ways is similar to that observed during general anaesthesia.

global workspace theory This theory proposes that mental contents (such as perceptions, thoughts, actions) become conscious when they gain access to a brain-wide 'global workspace', which allows them to be used flexibly in the control of behaviour.

integrated information theory (IIT) The IIT, a mathematical theory, proposes that conscious experiences arise from the integration of large quantities of information by the brain.

neural correlate(s) of consciousness (NCCs) Defined by Francis Crick and Christof Koch as 'the minimal neuronal mechanisms jointly sufficient for one specific percept'. The search for the NCC – or NCCs – is still the dominant approach in the neuroscience of consciousness.

neurotransmitters The chemical machinery of the brain. Neurotransmitters enable communication between neurons and synapses by crossing the divide ('synaptic cleft') between the axon of one neuron and the dendrite of another when a nerve impulse arrives. Neurotransmitters come in many varieties but can be broadly divided into excitatory and inhibitory subtypes.

prefrontal parietal network The network of regions comprising prefrontal and parietal cortical areas, which are involved in sensory integration, attention and higher-order cognitive functions. The prefrontal parietal network is frequently associated with consciousness, especially in so-called global workspace theories.

qualia A philosophical term broadly referring to the intrinsic properties of conscious experience – the redness of an evening sky, the sound of a bell, the warmth of a log fire and so on. Informally, it is simply 'the way things seem'.

rapid eye movement (REM) sleep A stage of sleep in which the eyes (though closed) move rapidly (hence the name). REM sleep occupies about one-quarter of a night's sleep and occurs mainly later on, towards the morning. Dreams are most commonly associated with REM sleep but they can occur during other sleep stages as well.

readiness potential A slowly rising electrical brain signal – detectable using EEG – that seems to precede voluntary decisions to perform actions. Controversy has raged over whether the existence of these potentials calls into doubt notions of 'free will'.

supplementary motor area Part of the frontal lobes and a possible neural source of the readiness potential. Direct electrical stimulation of this region produces the experiences of wanting to make a movement.

vegetative state A major disorder of consciousness that can follow severe brain injury. Patients in this condition seem awake but not aware. The vegetative state is called 'persistent' when it lasts for more than a year. The state is distinct from coma, in which patients show no signs of wakefulness or awareness, and from the so-called 'minimally conscious state', in which patients show brief and transient signs of awareness.

THE HARD PROBLEM

the 30-second neuroscience

Why are any physical processes, such as those happening in brains, ever accompanied by conscious experience? This is the 'hard problem', to be distinguished from the so-called 'easy problem' of explaining how the brain works. The hard problem has been with us for centuries – early 'dualist' thinkers, such as Descartes, divided the universe into 'mind stuff' and 'material stuff' – but it was only in the 1990s that the philosopher David Chalmers coined the term. Zombies – of the philosophical kind – also depend on the hard problem. A philosophical zombie (unlike the Hollywood type) is indistinguishable from a real person in all its behaviour, yet there is no conscious experience going on within. Not all philosophers believe in the hard problem. Daniel Dennett argues that consciousness is best defined in terms of the function that it supports and not by the raw essence of experience (so-called *qualia*). A challenge for this view is that no one is sure what consciousness is really for. Some people think the hard problem will prevent us from ever scientifically understanding consciousness, but this might be unfairly pessimistic. It could be that by solving the 'easy' problems of the brain, the hard problem will dissolve away.

RELATED BRAINPOWER
See also
NEURAL CORRELATES
OF CONSCIOUSNESS
page 82

CONSCIOUSNESS &
INTEGRATION
page 86

THE ANAESTHETIZED BRAIN
page 90

3-SECOND BIOGRAPHIES
RENÉ DESCARTES
1596–1650
Founder of dualism; discovered – or invented – the hard problem

DANIEL DENNETT
1942–
Famous for writing *Consciousness Explained*

DAVID CHALMERS
1966–
Coined the term 'hard problem'

30-SECOND TEXT
Anil Seth

Both have exemplary neural circuitry and regular income but only one has consciousness.

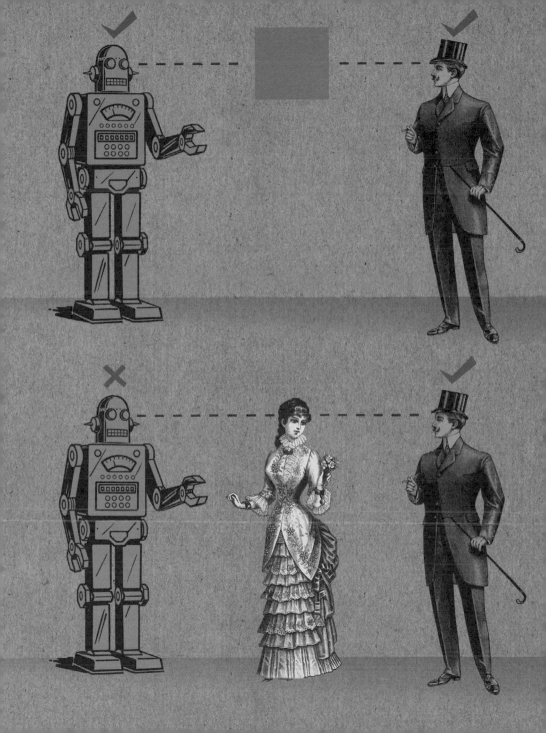

SLEEP & DREAMING

the 30-second neuroscience

3-SECOND BRAINWAVE
Sleep is the brain's way of
dealing with being awake.

3-MINUTE BRAINSTORM
Dream consciousness is
different from normal
consciousness. When
dreaming, we easily accept
bizarre events, show
reduced self-awareness and
generally lack experiences
of volition. This may have
to do with lower activity in
the prefrontal cortex during
dreaming. And dreaming is
not limited to REM. Sleep–
dream reports are also
common after waking from
early 'slow wave' sleep,
though these dreams are
relatively static, snapshot-
like images and usually lack
a 'self' character.

We spend about one-third of our lives sleeping and, when not dreaming, we are completely without consciousness. Even the humble fruit fly sleeps while some creatures, such as dolphins, sleep with half their brain at a time. Sleep matters: go one night without it and we suffer the next day; go too long and we would die. As we fall asleep, the rapid electrical activity found in normal waking fades away and slow, deep waves start coursing throughout the cortex. Most sensory input is blocked and nerve signals to muscles are interrupted, preventing us from acting out our dreams. Despite this, the brain remains almost as active as when we are awake. Sleep is usually divided into three stages of increasing depth, plus the 'rapid eye movement' (REM) stage in which brain activity is similar to waking and most dreams occur. There are many theories about why we sleep. Some researchers think sleep helps consolidate memories from the previous day. Others believe it rebalances neurotransmitter levels. Dreams are even more mysterious. Freud believed that they represent wish fulfilment of the subconscious. A more recent theory argues that they are just the brain making sense of its own activity when cut off from the world.

RELATED BRAINPOWER
See also
THE OSCILLATING BRAIN
page 46

THE ANAESTHETIZED BRAIN
page 90

THE IMAGINING BRAIN
page 122

3-SECOND BIOGRAPHIES
SIGMUND FREUD
1856–1939
Proposed that dreams represent
unfulfilled unconscious wishes

ALLEN HOBSON
1933–
An American psychiatrist
notable for his 'activation-input-
modulation' (AIM) theory of
dreaming

30-SECOND TEXT
Anil Seth

*Sleep gives your brain
the chance to get on
top of its paperwork,
chuckle over your day's
events and enjoy some
quality 'me' time.*

8 June 1916
Born Francis Harry
Compton Crick in
Weston Favell,
Northampton

1937
BSc in Physics from
University College
London

1939–45
Worked for the
Admiralty as one of
a team of scientists

1947
Began to study
biology; worked at the
Strangeways Research
Laboratory, Cambridge

1950
Research student at
Gonville & Caius College,
Cambridge

1952
Met and became friends
with James D. Watson

1953
Proposed the double
helical structure of DNA
with Watson

1954
PhD from Cambridge
on X-ray diffraction,
polypeptides and
proteins

1959
Fellow of the Royal
Society

1962
Shared Nobel Prize for
Physiology or Medicine
with James D. Watson
and Maurice Wilkins
(but not Rosalind
Franklin) for their
work on the
structure of DNA

1967
Published *Of Molecules
and Men*

1977
Left UK to work
full time at the
Salk Institute;
simultaneously held
a professorship at
the University of
San Diego

1981
Published *Life Itself:
Its Origin and Nature*

1982
Published paper with
Graeme Mitchison on the
function of REM sleep

1988
Published *What Mad
Pursuit: A Personal View
of Scientific Discovery*

1990
Began work with
Christof Koch on vision,
short- term memory and
consciousness

1994
Published *The
Astonishing Hypothesis:
The Scientific Search for
the Soul*

28 July 2004
Died in San Diego

FRANCIS CRICK

Perhaps best known as one

half of the Nobel-winning Crick and Watson partnership, the Lennon and McCartney of molecular biology, Francis Crick crammed enough into his illustrious career as a scientific theorist for a dozen great minds. Indeed, the strands of his interests are packed tightly together like the strands in the DNA helix. At the same time, huge upheavals, both intellectual and geographical, characterize his life, and his work is an exhilarating combination of hard work and inspired imaginative leaps, with occasional forays into more controversial areas, such as the panspermia theory and eugenics. The polar opposite of an ivory-tower academic, he believed that scientific theory could not be developed in isolation from the human experience. He was an energetic popularizer, writing several books explaining difficult concepts to eager but untrained minds.

Crick began academic life at University College, London, as a physicist, but in 1947 he switched to biology, studying for his PhD at Cambridge, before moving into genetics. After a 30-year career at Cambridge, he changed direction again, moved to the Salk Institute for Biological Research in La Jolla, California, taught himself neuroanatomy and focused his attention on theoretical neuroscience. In 1982, working with Graeme Mitchison, he produced a paper on

possible functions of REM sleep and then from the early 1990s focused his talents squarely on re-establishing the investigation of consciousness at the heart of neuroscience. Crick had been struck by the reluctance of neuroscientists to tackle this central problem, observing in a 1990 paper with Christof Koch that 'it is remarkable that most of the work in both cognitive science and the neurosciences makes no reference to consciousness'. His collaboration with Koch lasted until his death in 2004 and resulted in a series of influential theoretical articles about the relationship between brain activity and visual awareness, developing the notion of 'neural correlates of consciousness', or NCCs as they have come to be known.

Always a theorist rather than an experimenter, Crick's greatest gift was the ability to discern patterns and links, to take the wider view without losing the detail and to understand how different scientific disciplines need to work together to build a better idea. If scientific theory was a sport, Crick would have been a top coach, headhunting players from different disciplines to build a team that would achieve the desired result. His last book, *The Astonishing Hypothesis: The Scientific Search for the Soul*, vigorously promoted the idea that neurobiology had all the tools and skills it needed to work out the age-old question of why (and how) we are conscious.

NEURAL CORRELATES OF CONSCIOUSNESS

the 30-second neuroscience

3-SECOND BRAINWAVE
Brain activity can correlate with conscious experience, but does it explain it?

3-MINUTE BRAINSTORM
A big problem with the NCC approach is that it is difficult to ensure that the only thing that changes is the conscious experience. Usually, when we are conscious of 'X', we also say so, either verbally or by (for example) pushing a button. Thus, it is hard to disentangle the NCC from related brain processes, such as attention, memory and behavioural report.

A popular way to investigate

consciousness is to compare brain activity in unconscious and conscious conditions, either for different conscious states, such as sleeping or dreaming, or for different conscious experiences, such as seeing a house or a face. In binocular rivalry, one image (let's say a house) is shown to one eye and at the same time another (a face) is shown to the other eye. Because the brain cannot resolve this ambiguous sensory input into a single image, one's conscious experience alternates between the house and the face. Comparing brain activity between these two conditions should reveal the neural correlates of consciousness (NCCs) of the house (or face). Francis Crick (co-discoverer of the structure of DNA) and his colleague Christof Koch defined the NCC as the 'minimal neuronal mechanisms jointly sufficient for one specific percept'. Current experiments do not yet allow us to achieve such a detailed view. Showing that certain brain regions – or types of activity – correlate with consciousness does not show that they are sufficient, because other brain processes might be involved. Still, much has been learned, for example that consciousness is generally associated with activation of a large swathe of cortex involving prefrontal and parietal regions, and that 'top-down' connections from higher brain regions to sensory regions are also vital.

RELATED BRAINPOWER
See also
CONSCIOUSNESS & INTEGRATION
page 86

VOLITION, INTENTION & 'FREE WILL'
page 88

THE IMAGINING BRAIN
page 122

3-SECOND BIOGRAPHIES
CHRISTOF KOCH
1956–
A long-term collaborator of Crick and an early pioneer in the NCC approach with a focus on visual consciousness

GERAINT REES
1967–
A cognitive neuroscientist who has generated many insights into NCCs of visual consciousness; pioneered research looking into structural NCCs

30-SECOND TEXT
Anil Seth

Is it a face? Is it a house? Stop messing with my NCCs.

EMBODIED CONSCIOUSNESS

the 30-second neuroscience

3-SECOND BRAINWAVE
The experience of our bodily self and its location in space is actively constructed by the brain and is surprisingly open to change.

3-MINUTE BRAINSTORM
The brain's body model depends on multiple brain regions, including the parietal and somatosensory cortices, temporo-parietal junction and angular gyrus. Damage to these areas can lead to a variety of bizarre syndromes, including somatoparaphrenia (in which one denies ownership of a body part, sometimes attributing it to someone else) and xenomelia (the desire to amputate a completely healthy limb). Electrical stimulation and damage to the angular gyrus in patients can lead to out-of-body experiences.

Part of any conscious scene is

the experience of owning and identifying with a particular body. This seems so obvious we may take it for granted, but plenty of evidence shows that our body experience is actively constructed by the brain, just like our experience of the external world. In the now classic 'rubber hand illusion', synchronous stroking of a fake hand and a person's real hand, with visual attention focused on the fake hand, leads them to experience the fake hand as part of their body. This means that the brain infers which parts of the world belong to its body and which do not, on the basis of correlations between different senses. On the other hand (so to speak), if you are unlucky enough to lose a limb, you may continue to experience sensations emanating from that limb even though it no longer exists (so-called 'phantom limb syndrome'). This again shows that the brain builds a 'body model' that doesn't always match the physical body. Recent studies have taken this view even further. Using clever combinations of virtual reality, head-mounted cameras, multisensory stimulation and insights from 'out-of-body experiences', researchers have induced 'autoscopic' experiences that lead them to not only experience a fake hand, but an entire body – filmed or virtual – as their own.

RELATED BRAINPOWER
See also
THE BAYESIAN BRAIN
page 50

ALIEN HAND SYNDROME
page 112

3-SECOND BIOGRAPHIES
V. S. RAMACHANDRAN
1951–
Pioneer in investigation of phantom limb syndrome and developer of a novel low-tech 'mirror box' therapy

OLAF BLANKE
1969–
Known for his work on the neural basis of self-consciousness

THOMAS METZINGER
1958–
Philosopher and author of the 'self model theory of subjectivity', a theory of consciousness

30-SECOND TEXT
Anil Seth

Is this a hand I see before me? Is it mine? The rubber hand illusion tricks the brain into thinking the fake hand is real.

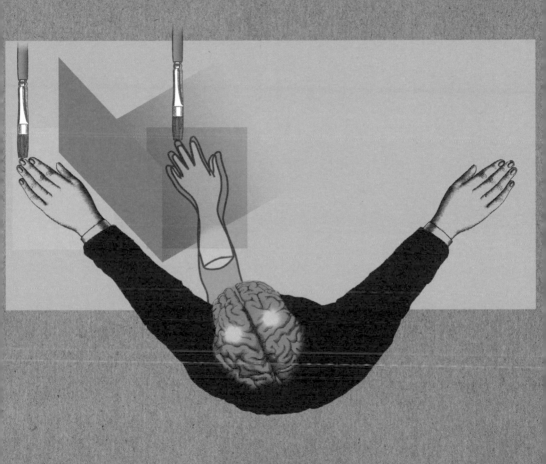

CONSCIOUSNESS & INTEGRATION

the 30-second neuroscience

3-SECOND BRAINWAVE
There is no consciousness 'hot spot' in the brain. Consciousness depends on the integration of neural activity among different brain regions.

3-MINUTE BRAINSTORM
Brain-imaging studies suggest that the prefrontal and parietal cortices are particularly important for consciousness and may form part of a global workspace. However, it is hard to know whether these regions generate consciousness itself, or whether they implement associated processes, such as attention, memory and verbal report of conscious experiences. The IIT is less specific about the underlying neuroanatomy, though it does stress the importance of interactions between thalamus and cortex.

Many neuroscientists believe that the critical processes for consciousness involve the integration of neural activity among different brain regions. According to 'global workspace theory', specific mental contents (such as perceptions, thoughts or intentions to act) remain unconscious unless or until they gain access to a 'global workspace', which broadcasts their contents across the brain, making them available for the flexible control of behaviour. This theory invites us to picture a theatre in which mental contents become conscious only when illuminated on a main stage where they can be seen by – and interact with – an audience. Integrated information theory (IIT) is also about networks, but its starting point is that every conscious experience is unique – one among a repertoire of possible experiences – resulting in the generation of an enormous amount of information. Consciousness is also integrated, in the sense that all the sounds, sights, thoughts and emotions we experience at any moment are bound together into a single conscious scene. IIT suggests this combination of information and integration can be quantified mathematically and this should correspond to the level of consciousness experienced. 'Integrated information' should be high during normal wakefulness and low during unconscious states, such as dreamless sleep.

RELATED BRAINPOWER
See also
THE HARD PROBLEM
page 76

NEURAL CORRELATES
OF CONSCIOUSNESS
page 82

THE ANAESTHETIZED BRAIN
page 90

3-SECOND BIOGRAPHIES
BERNARD BAARS
1946–
Psychologist, originator of global workspace theory and author of *A Cognitive Theory of Consciousness* (1988)

GIULIO TONONI
1960–
Neuroscientist and psychiatrist, originator of the integrated information theory of consciousness; also known for his work on the neurophysiology of sleep

30-SECOND TEXT
Anil Seth

Neural circuits are the dots on the page – consciousness is what happens when the music is played.

VOLITION, INTENTION & 'FREE WILL'

the 30-second neuroscience

3-SECOND BRAINWAVE
While there may be no such thing as 'free will', the experience of willing or intending behaviour certainly exists and can be localized within the brain.

3-MINUTE BRAINSTORM
Libet himself was uncomfortable with the implications of his own experiments. He proposed that consciousness might impose a 'veto' in between the onset of the readiness potential and the execution of the corresponding movement. This implies 'free won't' instead of 'free will'. But, whether it is 'will' or 'won't', the idea that consciousness can somehow intervene directly into brain processes is deeply problematic. Recent experiments have, therefore, looked for neural signatures of these conscious 'vetoes'.

In the 1980s, Benjamin Libet performed one of the most notorious experiments in modern neuroscience. He asked participants to raise one finger, at a time of their own choosing, and to notice, by looking at a rotating clock hand, when they consciously felt the urge to do so. While they did this, he recorded electrical activity in their brains, finding reliable patterns of activity – 'readiness potentials' – that preceded the time of the urge by about half a second. Some cite this as evidence that the brain commits to an action (raising a finger) before we ourselves are aware of the intention to do so, apparently challenging commonsense notions of 'free will'. However, many find Libet's results unsurprising – all events that depend on the brain, whether they are behaviours (raising a finger) or experiences (the conscious intention to do so), should have prior causes in the brain. The readiness potentials recorded by Libet are associated with a brain region called the 'presupplementary motor area'. Indeed, the neurosurgeon Itzhak Fried found that mild electrical stimulation of this area brings about an experience of intending to move, while stronger stimulation leads to actual movement. But the controversy rolls on even now. It seems that the idea of conscious free will is one we are determined to hang on to.

RELATED BRAINPOWER
See also
THE HARD PROBLEM
page 76

HOW WE PICK UP A
CUP OF COFFEE
page 108

ALIEN HAND SYNDROME
page 112

3-SECOND BIOGRAPHIES
BENJAMIN LIBET
1916–2007
Renowned for his original experiments on the timing of conscious intentions

PATRICK HAGGARD
1965–
Cognitive neuroscientist who has expanded on Libet's work in a variety of interesting ways, including looking for brain signatures of conscious vetoes

30-SECOND TEXT
Anil Seth

Who's pointing the finger here? You or the Mr Big that lives in your skull?

THE ANAESTHETIZED BRAIN

the 30-second neuroscience

3-SECOND BRAINWAVE
General anaesthesia taps into the on/off switch for consciousness – but is there just one switch, or many? We may not know how anaesthetics work, but it's a good thing that they do.

3-MINUTE BRAINSTORM
Unconsciousness should be distinguished from unresponsiveness. General anaesthetics can inhibit behavioural responses by acting on the brain stem and other paralyzing medicines are sometimes used in combination with them to prevent body reflexes during surgery. There are (rare) occasions when patients wake up during surgery and by relying on behavioural responses alone, we would not be able to tell. This is one reason why improved technologies for 'measuring consciousness' are worth developing.

Imagine a world without anaesthesia. Trips to the dentist would be one thing, but major surgery would be quite another. The development of general anaesthetics (GAs) – substances that reversibly induce full unconsciousness – has revolutionized medicine in the last century. It has also provided a powerful window into the brain basis of consciousness itself. We now know that a range of different substances can act as GAs, but we still don't really know how they work. What we do know is that the brain's electrical activity under deep anaesthesia is different to either waking or sleeping states, resembling more closely profound states of unconsciousness such as the vegetative state. Brain-imaging studies have shown that GAs affect many brain regions, including the parietal cortex and the thalamus, with weaker effects in sensory areas, such as the primary visual cortex. According to the 'thalamic switch' theory, GAs turn consciousness off by reducing activity in specific parts of the thalamus. However, it is not yet known whether this deactivation is a cause or a consequence of loss of consciousness. It could be that the thalamus is needed to allow other cortical areas to communicate and it is the loss of this communication that leads to unconsciousness.

RELATED BRAINPOWER
See also
THE BASIC ARCHITECTURE OF THE BRAIN
page 24

SLEEP & DREAMING
page 78

CONSCIOUSNESS & INTEGRATION
page 86

COMA & THE VEGETATIVE STATE
page 92

3-SECOND BIOGRAPHY
WILLIAM T. G. MORTON
1819–68
Pioneer of general anaesthesia; in 1846 he became the first to demonstrate the use of ether as an anaesthetic

30-SECOND TEXT
Anil Seth

*The off switch.
Root canal work
just wouldn't be
the same without it.*

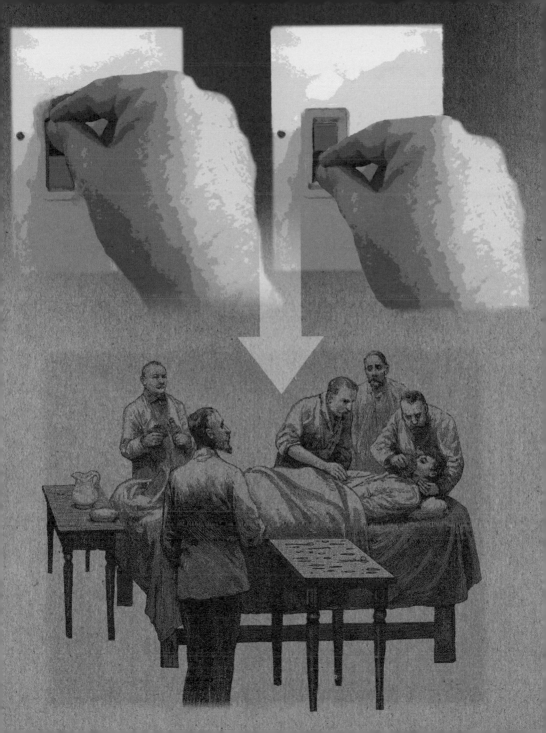

COMA & THE VEGETATIVE STATE

the 30-second neuroscience

3-SECOND BRAINWAVE
Being awake doesn't always mean being aware, and appearing unconscious on the outside doesn't always mean that consciousness is really gone.

3-MINUTE BRAINSTORM
Now that it is possible to communicate with some behaviourally unresponsive patients, what should we ask them? Asking whether they are in pain seems reasonable, but should we ask them if they want to stay alive? The ethics for all this has yet to be properly worked out. Also, fMRI scanners are too expensive and bulky to be widely used for these purposes, so recent research has explored the use of relatively cheap and portable EEG equivalents.

If you are unlucky enough to experience major brain damage, you might end up in a coma in which consciousness is absent and the brain's activity much diminished. If you survive, you might recover to a vegetative state in which you would seem to be *awake* but not *aware*. This condition can persist for years, even decades. These states are usually diagnosed by looking at patients' behaviour, but modern brain imaging has challenged many assumptions. In 2006, Adrian Owen and his colleagues asked an apparently vegetative patient to imagine either playing tennis or walking around their house while inside an fMRI scanner. Despite showing no external signs of understanding, the patient showed brain activity related to each task at the appropriate times. While these cases remain rare, they show that at least some patients previously thought to be unconscious are in fact conscious. The method can also be used to communicate with patients by asking them to imagine different activities for 'yes' and 'no', with answers read out on the scanner. Unfortunately, we are a long way from developing cures for these conditions. New techniques such as deep-brain stimulation are showing some promise, but only in a small minority of patients.

RELATED BRAINPOWER
See also
BRAIN IMAGING
page 58

THE ANAESTHETIZED BRAIN
page 90

3-SECOND BIOGRAPHIES
FRED PLUM
1924–2010
First to characterize the vegetative state

ADRIAN OWEN
1966–
Pioneer in the neuroscience of consciousness disorders, and originator of the house/tennis experiment

STEVEN LAUREYS
1968–
Pioneer in the brain imaging of vegetative and coma patients

30-SECOND TEXT
Anil Seth

Unawake does not mean unaware, even if you do have to communicate in hieroglyphs.

PERCEPTION & ACTION

area V4 V4 is one of the main subdivisions of the visual cortex, in the occipital lobes of the cerebral cortex. V4 was named by Semir Zeki in the 1970s and has been closely associated with the perception of colour, although it is not limited to this role.

attention A key cognitive and perceptual process in which the brain prioritizes the processing of some signals at the relative expense of others. There are many types of attention (such as attention to spatial locations, objects or object features). Attention is usually studied in the context of vision, but it is present for other senses as well.

cerebral cortex The deeply folded outer layers of the brain, which take up about two-thirds of its entire volume and are divided into left and right hemispheres that house the majority of the 'grey matter' (so called because of the lack of myelination that makes other parts of the brain seem white). The cerebral cortex is separated into lobes, each having different functions, including perception, thought, language, action and other 'higher' cognitive processes, such as decision making.

change blindness A phenomenon in which surprisingly large changes in a visual scene can go unnoticed if there is a gap between the images or if there are other more salient changes accompanying the transition. Change blindness, like inattentional blindness, suggests that our perception of the world may not be as reliable as we usually assume.

colour constancy The process by which our visual system compensates for changes in illumination that affect the balance of light waves reflected from coloured surfaces. For example, a bed of roses will result in a different mixture of wavelengths at dawn than at midday yet, even though colour depends precisely on this mixture, we will perceive the roses as a particular shade of red in both cases.

frontal lobes One of the four main divisions of the cerebral cortex and the most highly developed in humans compared to other animals. The frontal lobes (one for each hemisphere) house areas associated with decision making, planning, memory, voluntary action and personality.

inattentional blindness Related to change blindness, in inattentional blindness visible but unexpected objects can go unnoticed if attention is fully occupied elsewhere. In a famous demonstration by Dan Simons, volunteers failed to notice a man dressed as a gorilla walking across a basketball court when they were focusing on counting the number of successful passes.

motor cortex Part of the cerebral cortex, located towards the rear of the frontal lobes, which is responsible for the planning and execution of actions. The primary motor cortex sends control signals directly to the spinal cord and from there to the muscles. Higher-order motor areas are involved in more abstract sequencing and planning of actions and in initiating voluntary actions.

superior colliculus A pair of small structures beneath the cerebral cortex that play an important role in guiding fast visual reflexes, such as object tracking. Visual signals travelling via the superior colliculus do not reach the cortex.

WHY WE SEE COLOURS

the 30-second neuroscience

3-SECOND BRAINWAVE
The brain tries to figure out the true colour of a surface independent of the lighting conditions; so seeing colour isn't just detecting wavelengths of light.

3-MINUTE BRAINSTORM
There aren't seven colours in a rainbow. We just think there are because we have only a limited number of words for describing the colours we see. The way that we divide colours into categories affects the way we see them. For instance, some cultures don't have separate words for green and blue, and these people are worse at telling colours apart that are greeny blue (such as turquoise).

Our eyes are the mere starting point for seeing colours. The eyes contain special receptors for detecting the wavelength of light, this being the physical property of light most closely associated with colour. However, if we move an apple from sunlight to indoor candlelight, then we still see it as red even though the wavelengths of light reflecting off it can be very different due to the different lighting conditions. This aspect of colour is computed by the brain instead of the eyes. Moreover, some of the colours we can see (such as magenta) don't have a corresponding wavelength – they are entirely constructed by the brain. There is a part of the brain – 'area V4' – present in both hemispheres, that is responsible for the perception of colour. Damage to this region of the brain creates the experience of seeing the world in black and white. Why would there be a dedicated region of the brain for processing colour? The V4 region of the brain is thought to compute 'colour constancy' – that is, it infers the colour of a surface taking into account the lighting conditions. This ability may have evolved in our primate ancestors due to the need to reliably identify food sources, such as red fruit in green foliage.

RELATED BRAINPOWER
See also
THE BAYESIAN BRAIN
page 50

NEUROPSYCHOLOGY
page 56

SYNAESTHESIA
page 102

3-SECOND BIOGRAPHIES
EDWIN LAND
1909–91
Proposed one of the first accounts of colour constancy

SEMIR ZEKI
1940–
Famous for his studies on area V4

30-SECOND TEXT
Jamie Ward

Although the physical wavelength of light reflected from the green squares is the same, one of them is perceived by the brain as much brighter.

BLINDSIGHT

the 30-second neuroscience

3-SECOND BRAINWAVE
Some patients with damage
to the visual parts of the
brain claim not to be able to
see something but can then
accurately guess what they
are seeing. This is called
blindsight.

3-MINUTE BRAINSTORM
We all occasionally move
our eyes to something that
turns out to be important
without knowing why we
moved them there. It is
almost as if our eyes can see
something that 'we' cannot
yet see. However, this isn't
a discrepancy between the
eyes and the brain. Rather it
reflects two different routes
within the brain itself – a
fast seeing route (via the
superior colliculus) that
moves the eyes and a slow
seeing route (via the cortex)
than can detect the details
of what is seen.

Patient GY had a car accident as
a teenager that damaged a small region of his
brain and meant that he was 'blind' in one part of
his visual field. When shown a moving light in his
'blind' field, he would claim not to be able to see
it. When forced to guess whether the light moved
to the left or right he would be more than 90 per
cent accurate despite maintaining that he could
not see it. This paradox is termed 'blindsight'.
The key to understanding it comes from the fact
that there is not just one route that connects
the eyes to the brain but many (around ten routes
in humans), so damaging the brain might affect
one route selectively (whereas damaging the eyes
would tend to affect them all). These routes have
evolved for different functions – one uses light
to calibrate the biological clock (useful for getting
over jet lag) and another is used to orient the
eyes to sudden changes. These routes are still
functioning in patients such as GY, enabling
him to 'see' to some extent. The route that is
damaged in GY (and others like him) is the main
route involving the cortex – it is important for
seeing fine detail and is closely linked to the
conscious experience of seeing.

RELATED BRAINPOWER
See also
NEUROPSYCHOLOGY
page 56

NEURAL CORRELATES OF
CONSCIOUSNESS
page 82

MISSING THE OBVIOUS
page 106

3-SECOND BIOGRAPHIES
GEORGE RIDDOCH
1888–1947
Neurologist who studied visual
impairments in brain-injured
soldiers in the First World War

LARRY WEISKRANTZ
1926–
Researcher who coined the term
'blindsight'

NICHOLAS HUMPHREY
1943–
Researcher and philosopher with
an interest in vision

30-SECOND TEXT
Jamie Ward

*Blindsight means you
don't always get the
whole picture. You
may not have seen it,
but your brain saw it.*

SYNAESTHESIA

the 30-second neuroscience

3-SECOND BRAINWAVE
Our perceptions aren't
always triggered via our
sensory organs (eyes, ears,
etc.) but can be triggered by
activity in other regions of
the brain. For synaesthetes,
seeing colours could result
from hearing someone
speaking.

3-MINUTE BRAINSTORM
Daniel Tammett used his
synaesthesia to help him
remember the digits of pi to
over 20,000 decimal places.
For Daniel, each digit has its
own colour, size and shape
and the sequence of digits
is a colourful landscape
along which he can walk in
his mind's eye. For people
without synaesthesia,
visualizing things often
improves their memorability,
and linking sequences to
familiar routes can be used
as a strategy for learning,
say, the order of a pack of
playing cards.

For a small percentage of the
population, words may be coloured (so Tuesday
might be blue), music may have a taste or
numbers may be visualized on a twisting, turning
landscape. This remarkable way of experiencing
the world is termed synaesthesia. People with
synaesthesia have additional sensations (such as
someone smelling as well as hearing a sound) that
appear effortlessly to them. It emerges early in life
and tends to be stable; so if Tuesday is blue now,
then it will be tomorrow and next year. One key to
understanding this is to realize that our conscious
perceptions (such as of colour) are created by the
brain, so they can be turned on not only by our
sensory organs but also by activity in other parts
of the brain. In synaesthetes, the colour centre
of the brain may be turned on not only by seeing
colours but also by hearing words. The latter may
happen because people with synaesthesia have
unusual patterns of connectivity between regions
of the brain that tend to be more segregated in
others. This rewiring may be genetic (synaesthesia
runs in families) but it is not a disorder. For instance,
having synaesthesia may improve memory.

RELATED BRAINPOWER
See also
THE DEVELOPING BRAIN
page 28

NEURAL CORRELATES
OF CONSCIOUSNESS
page 82

WHY WE SEE COLOURS
page 98

3-SECOND BIOGRAPHIES
FRANCIS GALTON
1822–1911
Victorian polymath who
popularized the notion
of synaesthesia

V. S. RAMACHANDRAN
1951–
Neuroscientist who links
synaesthesia with evolution
of language and creativity

RICHARD CYTOWIC
1952–
Researcher and author of
The Man Who Tasted Shapes

30-SECOND TEXT
Jamie Ward

*Play me a blue note or
a hibiscus-scented one
or one that tastes of
cinnamon and honey.*

SENSORY SUBSTITUTION

the 30-second neuroscience

RELATED BRAINPOWER
See also
THE LOCALIZATION
OF FUNCTION
page 36

SYNAESTHESIA
page 102

TRAINING THE BRAIN
page 140

3-SECOND BRAINWAVE
Technology can be developed that translates vision into hearing or touch, providing blind people some access to the visual world. The brain translates the auditory or tactile input back into something like vision.

3-MINUTE BRAINSTORM
Is the success of sensory substitution dependent on the user having previous vision? Could a person, blind from birth, report new experiences as a result of these devices? There is no straightforward answer. Someone blind from birth would not fully understand what vision is like. People who became blind early in life have reported unusual experiences after using these devices (sensing something in front of them as their back is stimulated) but it is unclear if this is 'seeing' (they don't report shades of light and dark).

Sensory substitution devices

convert information from one sense into another. Their normal purpose is to enable blind people to 'see' by using their intact senses of hearing or touch. For instance, one early device used an array of pins on the wearer's back to create a two-dimensional tactile impression of a visual scene. This enabled blind people to recognize distant objects and even created a feeling of the objects being 'out there' in front of them, even though it was only their back that was stimulated. Modern versions are miniaturized and stimulate the user's tongue, being linked, via a computer, to a webcam on their head. Alternative devices use sounds to convey vision – for instance, different pixels in an image are translated to different pitches and different points in time. After training, users (blind or sighted) come to recognize the sound waves as 'shapes', and the sounds activate parts of the brain normally dedicated to visual or tactile shape detection. The ability of the brain to adapt to these devices is a striking example of plasticity that, in some very real sense, creates a cyborg – a functioning entity that is part man, part machine.

3-SECOND BIOGRAPHIES
PAUL BACH-Y-RITA
1934–2006
Pioneered this field from the 1960s with his tactile-visual devices

KEVIN O'REGAN
1948–
Scientist and philosopher who argues that vision can be experienced from other senses

PETER MEIJER
1961–
Inventor of an auditory sensory substitution device the 'vOICe'

30-SECOND TEXT
Jamie Ward

The resourceful brain can translate sound into a kind of vision, with a little help.

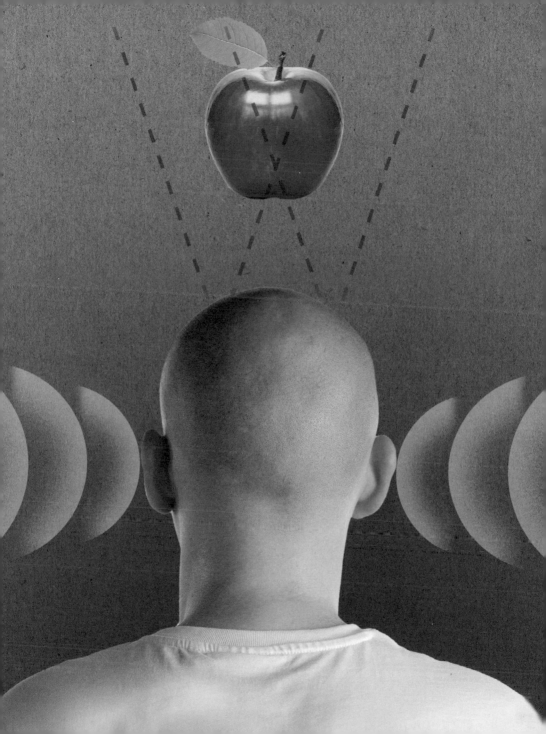

MISSING THE OBVIOUS

the 30-second neuroscience

3-SECOND BRAINWAVE

Not all objects in a crowded visual scene get processed fully. The brain contains a filtering mechanism (attention) that prioritizes a few things at a time.

3-MINUTE BRAINSTORM

There are similar examples in the domain of hearing. If we listen to several streams of conversation at the same time, we can, to some extent, filter out the irrelevant ones. We may even fail to notice when an unattended voice changes from male to female or English to German. In this example, unattended information, processed in the auditory parts of the brain, is not propagated to 'higher' regions of the brain (like the prefrontal and parietal areas) that allow It to be fully processed.

The visual world feels uniformly rich and detailed rather than patchy. However, our everyday experience of searching for a 'hidden' object – say, our car keys – that turns out to be right in front of us suggests that we don't process all the objects in our field of view to the same degree. There have been some fun experiments that illustrate this. If you are counting the number of passes in a basketball game, you may fail to notice a man in a gorilla suit who walks across the court (this is called inattentional blindness). Similarly, if a library assistant ducks below the counter to find your book and a new assistant pops up in the same place, then you may not notice the difference at all (this is termed change blindness). Vision and the rest of our senses are continually bombarded with information and the brain has developed a mechanism – attention – that acts as a filter by allowing some information to be prioritized (such as by amplifying its neural signal) at the expense of other information. This prevents us from being continually distracted (so we don't notice the touch of our clothing because we don't attend to that signal), but 'missing the obvious' can sometimes be the price that is paid.

RELATED BRAINPOWER

See also
NEURAL CORRELATES
OF CONSCIOUSNESS
page 82

BLINDSIGHT
page 100

HOW WE PICK UP A
CUP OF COFFEE
page 108

3-SECOND BIOGRAPHIES

DONALD BROADBENT
1926–93
Psychologist who conceptualized attention as a filter

RONALD RENSINK
1955–
Conducted pioneering studies on change blindness

30-SECOND TEXT
Jamie Ward

Pay attention. How many passes did you count? How many slam dunks? And why is the third player wearing that gorilla suit?

HOW WE PICK UP
A CUP OF COFFEE
the 30-second neuroscience

The skill of hand-eye

co-ordination is not as simple as it seems, as decades of research on robotics has found. Locating the cup of coffee is the first challenge. Our visual system can tell us where it is located on the retinal image, but we do not want to reach inside our eyes to find the coffee cup. To locate the coffee cup in the external world, we need to know the position of the eyes in the socket and the position of the head relative to the body (depth is yet another issue). So signals from the eye and neck muscles need to be linked to vision. In this way, we transform a visual image from something that is centred on the eyes to something that is centred on the body and, hence, can be acted upon. There appear to be separate pathways in the brain for locating and acting on an object (such as picking it up) versus knowing what the object is (such as recognizing it as a cup of coffee), although the two pathways normally work together effortlessly. Knowing what an object is affects how we interact with it – think picking up a paperweight versus an egg. Robots trained to pick up a paperweight will get sticky fingers when encountering the egg unless they have some kind of object recognition system.

RELATED BRAINPOWER
See also
THE BAYESIAN BRAIN
page 50

BLINDSIGHT
page 100

ALIEN HAND SYNDROME
page 112

3-SECOND BIOGRAPHIES
MEL GOODALE & DAVID MILNER
1943– & 1943–
Championed the idea that vision-for-action is different from vision-for-perception

RICHARD ANDERSEN
1950–
Neuroscientist who has explored how visual space is transformed from eye to body co-ordinates

30-SECOND TEXT
Jamie Ward

3-SECOND BRAINWAVE
Locating a cup of coffee involves linking visual information with information about the current posture of the body, which involves highly specialized mechanisms in the brain.

3-MINUTE BRAINSTORM
Computer vision and robotics can learn important lessons from the human brain. One lesson to be learned is that the body is important. We actively explore the world by moving our eyes, head and body, and this exploration gives us new sources of information. For instance, noticing how the visual world changes as a result of moving our head provides clues to depth and occlusion.

Brain the size of Jupiter, speaks 400 languages, can calculate pi to infinity, but still can't pick up a cup of coffee without spilling it.

9 July 1933
Born in London

1954
BA in physiology
and biology from The
Queen's College, Oxford

1966
Worked at Beth Abraham
Hospital, New York, with
survivors from the 1920s
encephalitis lethargica
pandemic

1966–2007
Instructor and later
Clinical Professor of
Neurology at Albert
Einstein College of
Medicine, New York

1973
Published *Awakenings*,
a report on the
experiences of his sleep
sickness patients

1985
Published *The Man Who
Mistook His Wife for a
Hat and Other Clinical
Tales*, his first bestseller

1992–2007
New York University
School of Medicine

1995
Published *An
Anthropologist on Mars:
Seven Paradoxical Tales*
about people with brain
disorders who also
manage creative
high-profile lives

2007
Joined the medical
faculty of Columbia
University Medical Center

2007
Published *Musicophilia:
Tales of Music and the
Brain*

2010
Published *The Mind's
Eye*, case studies about
the experience of visual
impairment

2012
Published *Hallucinations*,
studies of his own and
his patients' experience
of hallucinations, both
disease- and drug-induced

2012
Appointed Professor of
Neurology at the NYU
School of Medicine and
consulting neurologist
at its Epilepsy Center

2015
Died in Manhattan,
New York

OLIVER SACKS

Oliver Sacks was Professor of Neurology at the New York University School of Medicine. British born and a graduate of Oxford and New York universities, he lived and worked in the United States from the 1960s.

Sacks always strove to unite science with the arts, to find a sympathetic, universally comprehensible way to explain the behaviour of the brain and consciousness without neglecting the physiology and neural circuitry that underpins it. His bestselling books were based on case studies of his patients' and his own experiences (as a migraine sufferer and an experimenter with mind-altering drugs, for example). He presented the diseased, damaged or broken mind not through the clinical observations of an Olympian physician, but from the horse's mouth, in the words of the patients themselves, whether they were living with Tourette's, Parkinson's, aphasia, autism or other conditions. This offered the opportunity to report on the glories as well as the problems that could come from 'abnormal' responses to the world, an approach that accommodated both the pragmatic and the poetic. To the *New York Times*, Sacks was the Poet Laureate of Medicine, and in 2001 he was awarded the Lewis Thomas Prize for Writing about Science. His most famous book is probably *The Man Who Mistook His Wife for a Hat*, 24 essays reporting in from the extremes of the mind.

As a neurologist, Sacks was also known for the work he did in the 1960s with survivors from the 1920s pandemic of sleepy sickness (encephalitis lethargica), who had been comatose for almost 40 years. Sacks believed that the new experimental drug L-DOPA, although designed to improve dopamine uptake as a therapeutic intervention in Parkinson's disease, might wake them up; and it did, but the outcome was not universally happy. His second book *Awakenings*, is an unflinching account of what happened.

Sacks was showered with awards and his work continues to be invaluable in bringing difficult conditions to the attention of the everyday reader. However, he remains a somewhat controversial figure. A few fellow peers felt that he could have been exploiting his patients as material for his literary career; others, that he was possibly too idiosyncratic and insufficiently rigorous in his reporting methods. Yet in terms of public profile, he could be described as neurology's Stephen Hawking: his humane, empathetic, and engaging approach makes him the neurologist most likely to be recognized by non-neurologists.

ALIEN HAND SYNDROME

the 30-second neuroscience

Alien hand syndrome is found

after certain types of brain damage. Patients feel that movements of a limb (normally the arm) are not initiated by them and, hence, feel 'alien', as if the limb has 'a mind of its own'. The movements can be meaningful or meaningless. For instance, the alien hand might unzip a sweater that the other hand just zipped up, or the alien hand may levitate spontaneously, making 'tentacular' movements of the fingers. The character of Dr. Strangelove, in the movie of the same name, displayed this symptom. There are multiple regions of the brain involved in producing actions. The primary motor cortex sends signals down the spinal cord to produce movements of the limbs. However, the primary motor cortex itself receives signals from other parts of the brain in the frontal lobes that shape and control our actions. Normally, these parts of the brain function together – our actions don't feel alien because the parts of the brain that produce movements communicate with the parts of the brain that control and guide action. If our motor cortex becomes disconnected from these controlling regions, then movements are produced that are unexpected and, hence, 'alien'.

3-SECOND BRAINWAVE
Your arm might feel out of control if your motor cortex becomes disconnected from other parts of your brain.

3-MINUTE BRAINSTORM
The alien hand syndrome might have something in common with hallucinations. The former involves movement and the latter involves sensations (hearing voices or seeing faces that aren't there), so they are superficially different. However, they may reflect similar kinds of brain mechanisms, that is they reflect spontaneous activity within sensory or motor regions of the brain that are 'unexpected' by higher regions of the brain.

RELATED BRAINPOWER
See also
THE BAYESIAN BRAIN
page 50

VOLITION, INTENTION
& 'FREE WILL'
page 88

HOW WE PICK UP A
CUP OF COFFEE
page 108

THE SCHIZOPHRENIC BRAIN
page 150

3-SECOND BIOGRAPHIES
KURT GOLDSTEIN
1878–1965
Reported the first known case of alien hand syndrome in 1908

STANLEY KUBRICK
1928–99
Directed the movie *Dr. Strangelove* – how did he know about the symptom?

30-SECOND TEXT
Jamie Ward

It wasn't me, it was my alien hand that did it. I know it was your favourite mug. I'll get a mop and bucket.

COGNITION & EMOTION

amygdala A collection of bundles of neurons (nuclei) buried deep in the medial temporal lobes of the cerebral cortex, about the size and shape of a walnut. The amygdala are part of the limbic system and are involved in emotional processing and especially in the learning of emotionally salient associations. Aversive emotions, such as fear, are particularly dependent on the amygdala.

Broca's aphasia (and Wernicke's aphasia) Aphasias are disorders in the generation (Broca) or comprehension (Wernicke) of language and are associated with damage to different regions of the linguistic brain.

cerebral cortex The deeply folded outer layers of the brain, which take up about two-thirds of its entire volume and are divided into left and right hemispheres that house the majority of the 'grey matter' (so - called because of the lack of myelination that makes other parts of the brain seem white). The cerebral cortex is separated into lobes, each having different functions, including perception, thought, language, action and other 'higher' cognitive processes, such as decision making.

frontal lobes One of the four main divisions of the cerebral cortex and the most highly developed in humans compared to other animals. The frontal lobes (one for each hemisphere) house areas associated with decision making, planning, memory, voluntary action and personality.

hippocampus A seahorse-shaped area found deep within the temporal lobes. The hippocampus is associated with the formation and consolidation of memories and also supports spatial navigation. Damage to this area can lead to severe amnesia, especially for episodic (autobiographical) memories.

insular cortex Meaning 'island', the insular cortex is found at the bottom of a deep fold at the junction of the temporal, parietal and frontal lobes. It is involved in detecting and representing the internal state of the body (so-called 'interoception') and is increasingly associated with conscious emotional experiences and the sense of self.

introspection The act of observing or examining one's own mental states. Introspection is a key instance of metacognition.

limbic system An old-fashioned term relating to a collection of brain structures involved in emotion, motivation and memory. These include the amygdala, hippocampus, certain thalamic nuclei and specific regions of cortex.

neurons The cellular building blocks of the brain. Neurons carry out the brain's basic operations, taking inputs from other neurons via dendrites, and – depending on the pattern or strength of these inputs – either releasing or not releasing a nerve impulse as an output. Neurons come in different varieties but (almost) all have dendrites, a cell body (soma) and a single axon.

orbitofrontal cortex The part of the frontal lobes lying directly above and behind the eyes. The orbitofrontal cortex is involved in the processing of emotional and motivational information, particularly in relation to decision making.

prefrontal cortex The most frontal part of the frontal lobes, the prefrontal cortex is associated with high-level cognitive functions, such as metacognition, complex planning and decision making, memory and social interactions. Collectively, these operations are sometimes known as 'executive functions'.

somatic marker hypothesis The brainchild of Antonio Damasio, this theory emphasizes the role of emotions in decision making. It proposes that complex decisions rely on the perception of bodily states (somatic markers), which represent the emotional value of different options.

temporal lobes One of the four main divisions of the cerebral cortex. These lobes are found low to the side of each hemisphere and are heavily involved in object recognition, memory formation and storage, and language. The hippocampus is in the medial part of these lobes (the medial temporal lobe).

THE REMEMBERING BRAIN

the 30-second neuroscience

3-SECOND BRAINWAVE
Many different parts of the brain take on specialist memory functions, according to content (such as knowledge versus past events) or process (such as recollection versus familiarity).

3-MINUTE BRAINSTORM
The prefrontal cortex, regularly associated with complex thought, is also involved in all memory processes. This broad role probably reflects a need for humans to perform various sophisticated memory-based operations, such as generating and using strategies to retrieve faint memories and making assessments about what information is truly important to retain. Consequently, when we obey instructions to forget something, our prefrontal activity increases as our medial temporal lobe activity decreases.

In the 1950s, a young man, Henry Molaison, was experiencing severe epilepsy. The doctors decided to remove his medial temporal lobes, which were thought to be the source of his symptoms. This successfully treated his seizures, but there was an enormous price to pay. Although his short-term memory (his ability to retain information for a few seconds or minutes) was largely intact, he was unable to form new long-term memories. Consequently, his mind never moved beyond the 1950s and however frequently he visited a new person or place, they would always be unfamiliar. Studies of such patients established the way in which the medial temporal lobes (especially the hippocampus) turn short-term memories into permanent ones, with much of the rest of the temporal lobes acting as a long-term store. Even within these storage areas, there is further fragmentation, with semantic memories (such as what the capital of France is) located in a separate region (the temporal pole) from memories of past events (which are more widely distributed across the temporal cortex). Evidence also shows that different processes occur when we have a sense of vague familiarity about a past event, compared with when we can confidently recall it, and these are carried out by different parts of the medial temporal lobes.

RELATED BRAINPOWER
See also
THE LOCALIZATION OF FUNCTION
page 36

HEBBIAN LEARNING
page 38

NEUROPSYCHOLOGY
page 56

3-SECOND BIOGRAPHIES
BRENDA MILNER
1918–
The first scientist to study Henry Molaison

ENDEL TULVING
1927–
Pioneering researcher into long-term memory

30-SECOND TEXT
Daniel Bor

What is the capital of France? How do I make that chess piecemove? Your brain has a specific place for each memory function.

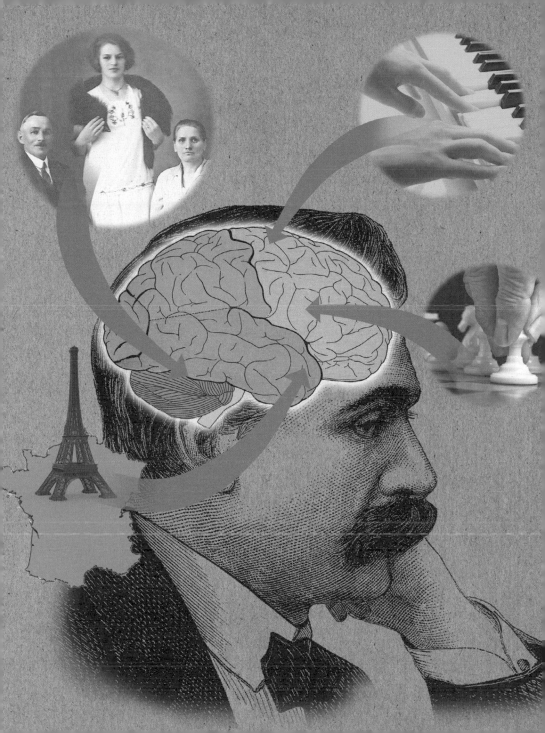

THE EMOTIONAL BRAIN

the 30-second neuroscience

Walking in the woods, you spy

a grizzly bear. Fear courses through your body and in an instant you are poised to flee. We might think that seeing the bear causes the feeling of fear, which leads to the charge of adrenaline that prepares us for flight (or fight, if you are brave). Things, however, are not so simple, as William James (and independently Carl Lange) surmised more than a century ago. They argued that an emotion (such as fear) is the result of perceiving changes in bodily state, not the cause of these changes. We feel afraid because our body is preparing to act in a particular way, not vice versa. Although controversial, this idea has stood the test of time; it is now widely accepted that emotions are deeply dependent on how the brain and body respond to each other, and the search is now on for the brain mechanisms involved. One key region is the amygdala, a collection of almond-shape structures buried deep inside the medial temporal lobes, which plays a key role in consolidating memories of emotional experiences. Damage to the amygdala can lead to dampening of emotional responses and especially the loss of fear. Many other neural areas are also involved, including the orbitofrontal cortex, which links emotion and decision making, and the insular cortex, which monitors the body's physiological condition.

3-SECOND BRAINWAVE
'I tremble, therefore I am afraid.' Emotion is the context-dependent perception of changes in the physiological condition of the body.

3-MINUTE BRAINSTORM
Emotion depends not just on bodily changes, but on the context of such changes. Stanley Schachter and Jerome Singer injected volunteers with adrenaline while an actor in the same room behaved either angrily or euphorically. The lucky ones experienced euphoria; the others, anger. Crucially, those who were informed about the physiologically arousing effects of adrenaline did not experience these emotions. The findings supported a 'two factor' theory, that a felt emotion depends on how the brain interprets changes happening in the body.

RELATED BRAINPOWER
See also
EMBODIED CONSCIOUSNESS
page 84

DECISION MAKING
page 130

3-SECOND BIOGRAPHIES
WILLIAM JAMES
1842–1910
With Carl Lange came up with the idea that emotions are perceptions of changes in bodily state

STANLEY SCHACHTER
1922–97
With Jerome Singer, conducted seminal experiments showing context dependency of emotional experience

ANTONIO DAMASIO
1944–
Reinvigorated the study of emotion with his 'somatic marker hypothesis'

30-SECOND TEXT
Anil Seth

You're not afraid of the bear. Your heart races and you are in a cold sweat because your body is setting up your fight/flight options.

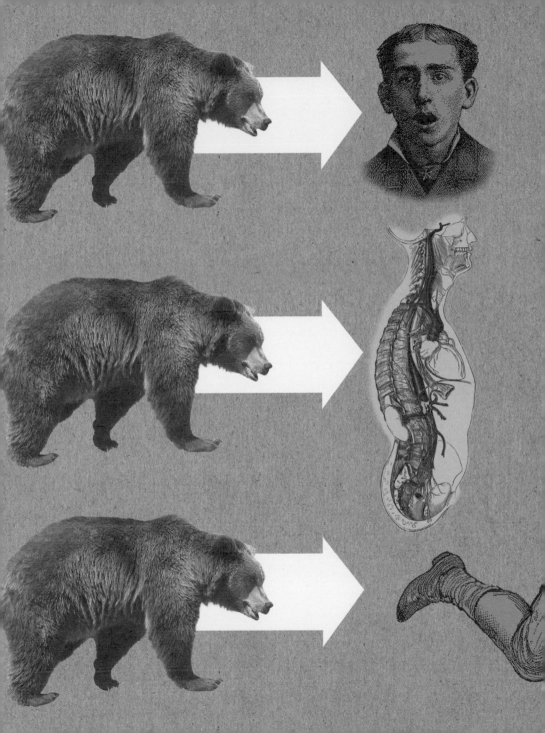

THE IMAGINING BRAIN

the 30-second neuroscience

3-SECOND BRAINWAVE
Imagination is a powerful, particularly human, skill. But instead of having specialized neural hardware, it is entirely reliant on our existing sensory regions.

3-MINUTE BRAINSTORM
Adrian Owen has revolutionized the study of vegetative state patients. He showed that some of these patients are wrongly diagnosed – they are completely paralyzed but still have an active inner consciousness. Owen tests this by asking questions in the fMRI scanner and the patient answers 'yes' by motor imagery (which activates the motor cortex) and 'no' by spatial imagery (activating the parahippocampus). Therefore, imagination gives these patients their only chance of a voice.

Einstein claimed that imagination is more important than knowledge. It is a strange mental skill that humans appear particularly adept at. For instance, if you close your eyes and imagine a black cat jumping up a green wooden fence, you will probably generate a surprising degree of detail. But is imagination a special process, distinct from perception, or is it intimately entwined with our senses? The evidence largely points to the latter. For instance, one patient, RV, had damage to the visual colour centres of his brain, causing a brain equivalent of colour blindness. RV was unable to work out what colour snow was when asked to imagine it. Another brain-damaged patient was unable to either perceive or imagine famous faces. Brain-scanning studies have dramatically extended this view, for example, by showing that auditory imagery activates the auditory cortex, similar to if the same sounds were actually heard. And bringing to mind fearful imagery is sufficient to activate the fear centre of the brain, the amygdala. The clear, and evolutionarily efficient, picture is that any part of our brains responsible for direct experience can also be co-opted to aid our imagination.

RELATED BRAINPOWER
See also
THE BAYESIAN BRAIN
page 50

NEUROPSYCHOLOGY
page 56

WHY WE SEE COLOURS
page 98

3-SECOND BIOGRAPHIES
STEPHEN KOSSLYN
1948–
Cognitive neuroscientist known for his work on mental imagery. He argues that imagination is composed of several different sub-processes

MARTHA FARAH
1955–
Neuropsychologist, who was one of the first to document patients lacking both perception and imagination in specific domains

30-SECOND TEXT
Daniel Bor

Just imagine what I could imagine if I had lived a fuller life and gained some more exciting data to play with.

28 June 1824
Born in Sainte-Foy-la-Grande, Gironde, France

1844
Graduated from medical school at Hotel Dieu, Paris

1848
Became Prosector of Anatomy at the University of Paris Medical School (the individual who makes dissections for anatomy students), and secretary of the Anatomical Society

1848
Founded a pro-Darwinist society of free thinkers

1849
Performed the first surgery in Europe using hypnotism as anaesthesia

1853
Professor of Surgery at the University of Paris

1856
Published *Aneurysms and their Treatment*

1859
Founded the Anthropological Society of Paris; published *The Ethnology of France*

1861
Performed autopsy on M. Leborgne, demonstrated that he has lesions on the frontal cortex of the left hemisphere, locating the area of the brain governing articulate speech

1865
Published *General Instructions on Anthropological Research*

1867–68
Elected to the chair of External Pathology at the University of Paris Faculty of Medicine, then Professor of Clinical Surgery

1872
Founded *The Anthropological Review*

1875
Published *Instructions on Craniology and Craniometry*

1876
Founded the School of Anthropology

9 July 1880
Died in Paris

PAUL BROCA

A child prodigy (he graduated when he was 16 and gained his medical degree at 20), Paul Broca was a physician, surgeon and anatomist who spent all of his student years and working life in Paris. His free-thinking, pro-Darwin stance and his interest in anthropology (he founded The Anthropological Society of Paris in 1859) got him in trouble with the Church and state authorities, yet they did not prevent him from leading a successful public life; he was elected to the French Senate, became a member of the Académie de Médecine and received the Légion d'Honneur.

Although he made many important contributions to other fields of medicine (including cancer, infant mortality and public health) it is as a neuroanatomist that he is best remembered. Broca was the first to describe the evidence of trepanning found in Neolithic skulls and made great advances in cranial anthropometry and comparative anatomy of the brain, providing invaluable data on its weight and size.

Broca is best known today as the first to produce reliable, replicable physical evidence of the localization of function in the brain, laying the ground for future research into brain lateralization. Franz-Joseph Gall (who died when Broca was four) had championed the idea of phrenology, the theory that different parts of the brain give rise to different actions emotions and moods. He was a theorist, making inferences from the cranium instead of the brain within. In 1861, after attending a lecture by Ernest Aubertin, a supporter of the idea of localized function, especially of speech, Broca was inspired to look for evidence within. As Professor of Clinical Surgery at the University of Paris, he had access to several hospitals, and in one of them he found M. Leborgne, an aphasic patient who could not articulate speech (although his comprehension was unimpaired). When Leborgne died, Broca performed an autopsy and discovered lesions on the frontal lobe of his brain's left hemisphere. Further autopsies on a statistically significant number of similar patients replicated the result. This was the first documented evidence that brain functions were localized and that the two hemispheres operated differently to each other (the right hemispheres on the autopsied bodies were all unblemished). The area was named after Broca by the Scottish neurologist David Ferrier.

Subsequent MRI scans that have been carried out on Leborgne's brain (preserved in the Museum of Mankind in Paris) indicate that there may be more to it than lesions, but they do not diminish Broca's contribution to neuroanatomy.

THE LINGUISTIC BRAIN

the 30-second neuroscience

The human brain has the unique capacity to use language to describe novel situations. The relationships between symbols and their meanings are learned within specific cultural contexts and are not inherited like other biological traits, such as eye colour. Yet the human brain appears to have a predisposition to learn language. This innate ability is supported by specialized brain regions developed in humans. 'Broca's area', in the inferior frontal gyrus, is thought to play a central role in processing syntax, grammar and sentence structure. Patients with damage to this region show symptoms of expressive aphasia (also known as Broca's aphasia), which is characterized by an inability to produce fluent, grammatical sentences. On the other hand, 'Wernicke's area', situated in the superior temporal gyrus, is involved in the comprehension of language. Damage to this region results in deficits in understanding written and spoken language, a symptom called receptive aphasia (also known as Wernicke's aphasia). These language areas are directly connected to each other via fibres – called the arcuate fasciculus – and comprise the core of the linguistic brain, predominantly in the left hemisphere.

RELATED BRAINPOWER
See also
THE LOCALIZATION OF FUNCTION
page 36

NEUROPSYCHOLOGY
page 56

LEFT BRAIN VS RIGHT BRAIN
page 68

3-SECOND BIOGRAPHIES
PAUL BROCA
1824–80
French anatomist and anthropologist

CARL WERNICKE
1848–1905
German neurologist who found lesions to superior temporal gyrus results in deficits in understanding of language

30-SECOND TEXT
Ryota Kanai

Broca's area is the same in male and female brains. Miscommunication between the sexes takes place in another part of the brain entirely.

3-SECOND BRAINWAVE
The comprehension and production of human language are processed by distinct brain regions within the left hemisphere.

3-MINUTE BRAINSTORM
The first few years in life are critical for learning a first language. Feral children without exposure to language fail to develop full linguistic abilities. They can learn many words, but their syntax never reaches a normal level. Second languages learned during the critical period are processed in the same regions of Broca's area and Wernicke's area as the first language, while different regions of Broca's area are used for a second language learned after puberty.

METACOGNITION

the 30-second neuroscience

3-SECOND BRAINWAVE
Metacognition refers to the awareness of one's own thought, memory, experience and action as a result of introspective interrogation. It is, literally, cognition about cognition.

3-MINUTE BRAINSTORM
Can animals introspect? What are the differences between the brains of animals that are capable of metacognition and those that aren't? One way to test for evidence of metacognition in animals is to examine whether they can adjust their behaviour based on the reliability of their decision. Some – such as macaque monkeys, dolphins and rats – exhibit signs of metacognition when performing a perceptual or memory task. Others, such as pigeons, do not.

How often do you think about thinking? The human brain does not simply convert sensory signals into the execution of actions, but can also assess the quality of its own perceptual experiences, interrogate the reliability of memory, and monitor the results of its own actions. These abilities to access internal mental states through introspection are called 'metacognition'. We make use of this metacognitive capacity spontaneously in everyday life. One example is when we evaluate our confidence when making a choice. If you are taking an exam, you might be confident in answering some of the questions, but less confident with others. Metacognition is not only important for monitoring how we learn new information, but also for communicating our subjective experiences with others. When you decide what to order in a restaurant, you spontaneously reflect upon your decision by thinking about the experience of eating. In scientific experiments, metacognition is often used as a test of the presence of conscious perception or explicit memory, because we cannot introspect on unconsciously processed information. Current research points to the anterior part of prefrontal cortex – a region particularly expanded in humans through evolution – as key for metacognitive processing.

RELATED BRAINPOWER
See also
DECISION MAKING
page 130

MIRROR NEURONS
page 132

THE MEDITATING BRAIN
page 152

3-SECOND BIOGRAPHY
JOHN FLAVELL
1928–
American developmental psychologist who established metacognition as a research area

30-SECOND TEXT
Ryota Kanai

I think therefore I think. This seems to be a particularly human ability. Rodin captured one in flagrante cogito.

DECISION MAKING
the 30-second neuroscience

3-SECOND BRAINWAVE
Feelings provide the basis for human reason – brain-damaged patients left devoid of emotion struggle to make the most elementary decisions.

3-MINUTE BRAINSTORM
Although we need emotions to make decisions, their input means we're not the cold rational agents that traditional economics assumes us to be. For instance, Daniel Kahneman demonstrated with Amos Tversky that the negative emotional impact of losses is twice as intense as the positive effect of gains, which affects our decision making in predictable ways. For one thing, it explains our stubborn reluctance to write off bad investments.

From Plato's charioteer of reason controlling the horse of passion, to Freud's instinctual id suppressed by the ego, there's a long tradition of seeing reason and emotion as being in opposition to one another. Translating this perspective to neuroscience, one might imagine that successful decision making depends on the rational frontal lobes controlling the animalistic instincts arising from emotional brain regions that evolved earlier (including the limbic system, found deeper in the brain). But the truth is different: effective decision making is not possible without the motivation and meaning provided by emotional input. Consider Antonio Damasio's patient, 'Elliott'. Previously a successful businessman, Elliott underwent neurosurgery for a tumour and lost a part of his brain – the orbitofrontal cortex – that connects the frontal lobes with the emotions. He became a real-life Mr Spock, devoid of emotion. But instead of this making him perfectly rational, he became paralyzed by every decision in life. Damasio later developed the somatic marker hypothesis to describe how visceral emotion supports our decisions. For instance, he showed in a card game that people's fingers sweat prior to picking up from a losing pile, even before they recognize at a conscious level that they've made a bad choice.

RELATED BRAINPOWER
See also
VOLITION, INTENTION & 'FREE WILL'
page 88

MISSING THE OBVIOUS
page 106

THE EMOTIONAL BRAIN
page 120

3-SECOND BIOGRAPHIES
DANIEL KAHNEMAN
1934–
Pioneer in the psychology of decision making; published a bestselling popular book about his research in 2011, *Thinking Fast and Slow*

ANTONIO DAMASIO
1944–
Neurologist, author and researcher, based at the University of Southern California

30-SECOND TEXT
Christian Jarrett

Stick or twist? How can I make a rational decision with those two bozos using my frontal lobes as a wrestling dojo?

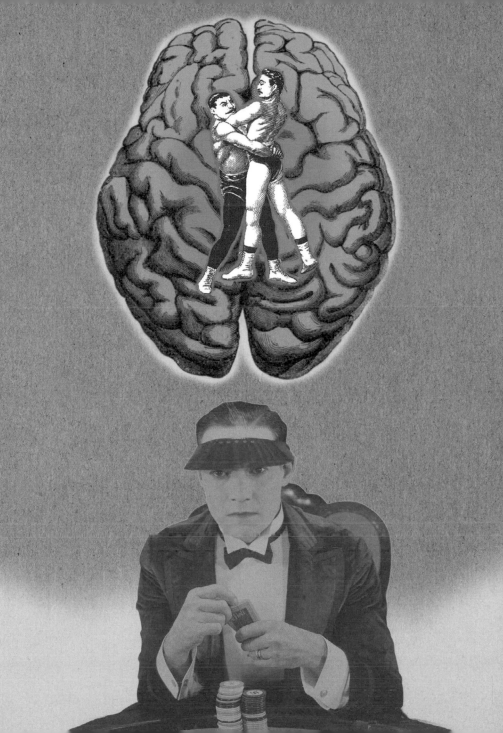

MIRROR NEURONS

the 30-second neuroscience

Mirror neurons were discovered by chance in the 1990s during research conducted in Giacomo Rizzolatti's laboratory at the University of Parma, Italy. Rizzolatti's team had been recording the electrical activity of motor neurons in the front of monkeys' brains, cells known to be involved in the planning and execution of bodily movements. The revelation came when one of the researchers reached for some of the raisins used as treats for the monkeys. To the scientists' astonishment, they realized that the monkeys' motor cells had fired, as if they had made the same movement as the researcher. In other words, these cells seemed to have mirror-like properties – they were activated during the execution of an action and by the sight of someone else performing that action. For years the race was on to confirm whether humans have mirror neurons, too – this is no easy task because recording from individual neurons in humans is usually too invasive. However, in 2010, a team led by Roy Mukamel was able to record from hundreds of neurons in the brains of epilepsy patients. The researchers identified a subset of mirror-like cells in the frontal cortex that responded both when the patients performed a given hand gesture or facial expression and when they watched a video of someone else performing those actions.

3-SECOND BRAINWAVE
Mirror neurons fire when you perform an action or you see someone else perform that same action.

3-MINUTE BRAINSTORM
The discovery of mirror neurons caused huge excitement in the field of neuroscience and beyond, largely because some experts claimed mirror neurons to be the source of human empathy. However, this is disputed. One obvious objection is that we're quite clearly capable of understanding actions, such as slithering and flying, that we are unable to perform ourselves. The suggestion that autism is caused by a 'broken' mirror neuron system also lacks scientific support.

RELATED BRAINPOWER
See also
NEURONS & GLIAL CELLS
page 16

HOW WE PICK UP A CUP
OF COFFEE
page 108

THE IMAGINING BRAIN
page 122

3-SECOND BIOGRAPHIES
GIACOMO RIZZOLATTI
1937–
Lead researcher at the University of Parma, where mirror neurons were first discovered

V. S. RAMACHANDRAN
1951–
Influential neuroscientist and author, renowned for his bold claims about mirror neurons (including saying that they will do for psychology what DNA did for biology)

30-SECOND TEXT
Christian Jarrett

It's not fair. My neurons have worked just as hard as John's, but I don't see any sweet raisins tumbling into my bowl.

THE CHANGING BRAIN

amygdala A collection of bundles of neurons (nuclei) buried deep in the medial temporal lobes of the cerebral cortex, about the size and shape of a walnut. The amygdala are part of the limbic system and are involved in emotional processing and especially in the learning of emotionally salient associations. Aversive emotions, such as fear, are particularly dependent on the amygdala.

comparator theory An influential theory about the cognitive dysfunctions underlying schizophrenia. Introduced by Irwin Feinberg and substantially developed by Chris Frith, it proposes that delusions – especially delusions of control – originate in failures to distinguish properly between self-generated and externally caused sensations.

dementia The loss of cognitive ability to a point where it impairs the ability of a person to function. Memory loss is a defining feature but there are other impairments as well. There are many forms of dementia, of which the most well known is Alzheimer's disease. Most instances of dementia are due to the degeneration of neural networks in the brain and are usually irreversible.

hippocampus A seahorse-shaped area found deep within the temporal lobes of the brain. This area is associated with the formation and consolidation of memories and also supports spatial navigation. Damage to the hippocampus can lead to severe amnesia, especially for episodic (autobiographical) memories.

olfactory system One of the most evolutionarily ancient parts of the brain. The olfactory system underpins the sense of smell and is less well-developed in humans than in many other animals. Signals from olfactory sensory neurons in the nose are conveyed to the olfactory bulb deep inside the brain. Olfaction and taste are distinct from the other senses in responding to chemical stimulation.

pheromones Chemical signals secreted by animals, which act as signals to others of the same species. Pheromones serve multiple purposes in social animals, including signalling alarm, fostering aggregation and helping navigation by marking out paths.

prefrontal cortex The most frontal part of the frontal lobes, the prefrontal cortex is associated with high-level cognitive functions, such as metacognition, complex planning and decision making, memory and social interactions. Collectively these operations are sometimes known as 'executive functions'.

proprioceptive system Proprioception refers to the sense of the position of the various parts of the body and is distinct from both exteroception (the classical senses directed at the outside world) and interoception (the sense of the internal bodily state). Like other sensory pathways, the proprioceptive system involves a pathway from the sensory periphery through the thalamus and to dedicated parts of the cortex.

spatial memory A particular type of memory involving information about one's location and orientation, necessary for finding one's way around. Spatial memory depends on the hippocampus, in the medial temporal lobes. In rats, specific hippocampal neurons – place cells – activate only when the rat is in a particular place in its environment, giving rise to the notion of a 'cognitive map'.

synapses The junctions between neurons, linking the axon of one to a dendrite of another. Synapses make sure that neurons are physically separate from each other so that the brain is not one continuous mesh. Communication across synapses can happen either chemically via neurotransmitters or electrically.

NEUROGENESIS & NEUROPLASTICITY

the 30-second neuroscience

3-SECOND BRAINWAVE

Neurogenesis and neuroplasticity allow the brain to adapt to the different demands placed on it at different stages of life.

3-MINUTE BRAINSTORM

Dominant male mice emit airborne chemical signals (pheromones) that stimulate neurogenesis in the olfactory bulb and hippocampus in the brain of female mice. The new neurons in the female's brain influence her choice of mate, causing her to show a strong preference to mate with dominant males. Whether something similar occurs in the human brain remains to be seen. In man as in mouse, however, the olfactory bulb and hippocampus are sites of significant adult neurogenesis.

Neurogenesis populates the

growing brain with neurons. Neuroplasticity adapts neurons and networks to the changing sensory environment. For much of the twentieth century, the belief was that neurogenesis only takes place before birth and through early childhood, after which the brain's structure was fixed. Today, we know that the brain is modified throughout our lives by neurogenesis. This changes the brain's wiring diagram because the new neurons form new synapses that must be incorporated into existing networks. However, how neurogenesis is stimulated and its functional significance remains poorly understood. Neurogenesis in the adult hippocampus, a centre for the formation of spatial memory, is one exception. Here, neurogenesis may be stimulated by learning one's way in a new environment. Once incorporated, the new hippocampal neurons and their synapses contribute to spatial memory functions and even cause the hippocampus to grow. The finding that taxi drivers have an enlarged hippocampus shows this link between exercising a brain function and growth of the region supporting that function. It is reassuring to know that the brain remains responsive and changeable throughout life. Exercise it and you can look forward to becoming both older (inevitably) *and* wiser (electively).

RELATED BRAINPOWER

See also
NEURONS & GLIAL CELLS
page 16

THE DEVELOPING BRAIN
page 28

THE REMEMBERING BRAIN
page 118

THE AGEING BRAIN
page 144

3-SECOND BIOGRAPHY
JOSEPH ALTMAN
1925–2016
First to report neurogenesis in the adult mammalian brain in the 1960s, though his work was largely ignored at the time

30-SECOND TEXT
Michael O'Shea

The hippocampus, the brain's GPS, can update and expand its route-finding program throughout adult life; good news for London's cab drivers.

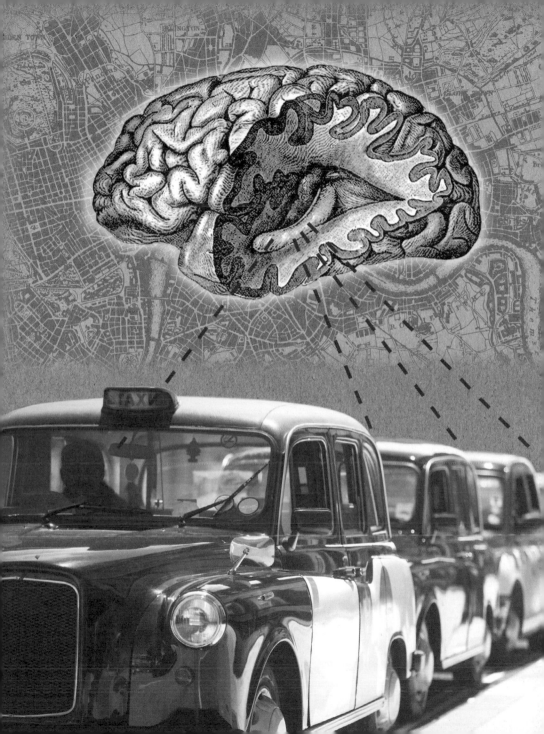

TRAINING THE BRAIN

the 30-second neuroscience

3-SECOND BRAINWAVE
Keeping your brain trim and smart by mental exercise sounds sensible, but it isn't well supported by the scientific evidence – at least not yet.

3-MINUTE BRAINSTORM
Although normal adults barely benefit from brain training, there is tentative evidence that this practice aids a range of disorders, including ADHD, early Alzheimer's disease and even schizophrenia. Why brain training works for these clinical populations is unclear, but it may be that some clinical symptoms arise from a particularly low working-memory capacity and brain training returns this to near normal levels, thus alleviating the more specific problems.

The premise behind brain training is that the brain is just like a muscle and that regular mental exercises can lead to general cognitive performance gains, as well as protection from atrophy. It sounds intuitive, but there is as yet almost no science to support this assumption. In fact, there is good evidence that brain training is pointless for most. For instance, one large-scale online study by Adrian Owen involved more than 11,000 participants between the ages of 18 and 60, who trained on various standard brain-training memory and reasoning tasks for six weeks. Although performance naturally improved on the tasks being trained, crucially there were no improvements on similar, but untrained tasks. This is probably because twenty-first century adults lead complicated enough lives and so we are constantly 'trained' by the need to understand our PCs, smartphones and computer games, not to mention the sudoku and crossword puzzles that so many of us enjoy. One recent lab-based training paradigm that has shown some promise, however, involves the tricky task of keeping in mind two different streams of information · simultaneously. Not only did performance increase dramatically over the weeks of training on this fiendish exercise, but so did IQ, particularly for those who started in the lower IQ range.

RELATED BRAINPOWER
See also
NEUROGENESIS & NEUROPLASTICITY
page 138

THE AGEING BRAIN
page 144

THE SCHIZOPHRENIC BRAIN
page 150

THE MEDITATING BRAIN
page 152

3-SECOND BIOGRAPHY
ADRIAN OWEN
1966–
Demonstrated the limited use of standard brain training techniques

30-SECOND TEXT
Daniel Bor

A daily crossword workout may make you a champion cruciverbalist, but it won't help you with quantum physics.

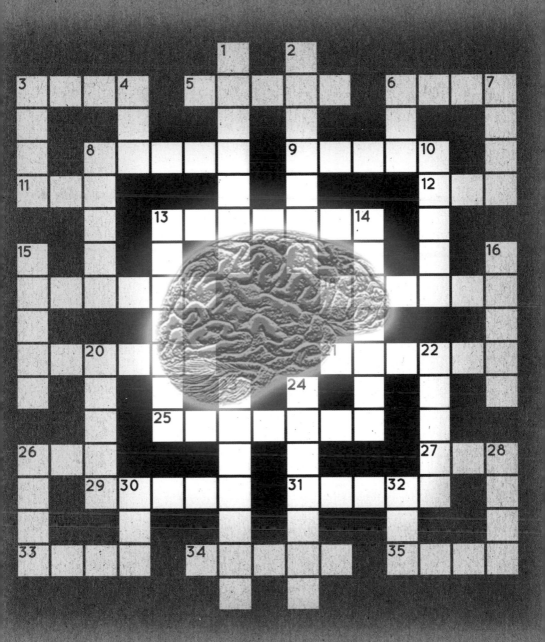

THE BRAIN'S PERSONALITY

the 30-second neuroscience

3-SECOND BRAINWAVE
You are your brain. Your personality emerges from the interaction of different networks in the brain, shaped by your genes and personal history.

3-MINUTE BRAINSTORM
One way to measure individual differences in brain structure relevant to personality is called 'voxel-based morphology', or VBM. VBM quantifies fine differences in brain volume between individuals, which can then be related to different facets of personality. For example, a recent study showed that brain volume in the 'posterior superior temporal sulcus' (pSTS) predicted the level of loneliness in a sample of people. Not yet an 'explanation' of loneliness, but the pSTS is linked to the processing of social cues, which seems very relevant.

Who are you? You are different from other people – even genetically indistinguishable twins are not born identical. A complex network of environmental, genetic and developmental influences combines to shape the various parts that make up your body – with your brain being especially susceptible. These influences can even occur before birth – during pregnancy, a modification of the womb composition by alcohol, drugs or changes in diet can lead to major changes in behaviour and personality later in life. During normal development, people – and their brains – become receptive to a wide range of environmental influences, including, critically, those arising from relationships with other humans. These differences in brain development coincide with changes in the wiring of the neural networks and these changes lie at the heart of personality differences that neuroscience is just beginning to unravel. So far, the evidence suggests that how neurotic, extrovert or intelligent we are seems to be associated with size and shape of our brains and the activity in the different brain areas. For example, it is well known that the amygdala, a nut-sized brain structure, is a key player in processing fear. Interestingly, it is hyperactive in anxious people or the very phobic, and the higher the anxiety of the patient, the higher the amygdala activity when they see fearful faces.

RELATED BRAINPOWER
See also
THE DEVELOPING BRAIN
page 28

BRAIN IMAGING
page 58

TRAINING THE BRAIN
page 140

THE AGEING BRAIN
page 144

3-SECOND BIOGRAPHIES
HANS EYSENCK
1916–97
Developed a theoretical framework about personality and the brain

RYOTA KANAI
1977–
Pioneer, with Geraint Rees, of linking neuroimaging to individual differences using VBM

30-SECOND TEXT
Tristan Bekinschtein

Macbeth was wrong. We are learning that there is, after all, an art to find the mind's construction in the face. Sorry Shakespeare.

THE AGEING BRAIN

the 30-second neuroscience

3-SECOND BRAINWAVE
One drawback of such a bright brain is its decline in later years, when the cortex thins and we become increasingly susceptible to dementia.

3-MINUTE BRAINSTORM
One ray of hope comes from recent evidence to overturn the view that neurogenesis, the formation of new neurons, does not occur in adulthood. Neurogenesis has been observed in the olfactory bulb, responsible for smell, and the hippocampus, a critical region for memory formation. However, like everything else, neurogenesis rates slow as we age. A key question for future research is whether this process can be harnessed to fight Alzheimer's and other age-related brain diseases.

If you're getting on in years,

you might be complaining to your more sprightly friends and family, as you mix up their names, that your brain isn't what it once was. Unfortunately, this is one example where science heavily reinforces our intuitions that the general ageing process is at least as brutal on our brains as the rest of our bodies. We start our lives after birth with a full complement of neurons, but the connections between them explode in number in the first 15 months or so of life and continue to sprout aggressively until we end our teenage years. However, very soon after this, in many ways our brain has already reached its peak and the only direction is down. Although certain brain regions decline faster than others, we lose on average approximately 10 per cent of our grey and white matter every decade of our adult lives. Mirroring this, our powers of reasoning, as measured by non-verbal IQ tests, peaks in our early 20s and declines steadily after this. And if that weren't depressing enough, our long-lived brains end up in our last decades being particularly susceptible to various forms of disease, with Alzheimer's disease the main concern. After the age of 65, Alzheimer's is increasingly common and more than 40 per cent of those over 85 have it.

RELATED BRAINPOWER
See also
THE DEVELOPING BRAIN
page 28

NEUROGENESIS &
NEUROPLASTICITY
page 138

3-SECOND BIOGRAPHIES
ALOIS ALZHEIMER
1864–1915
Neuropathologist who discovered Alzheimer's disease

JOHN MORRISON
1952–
Prominent modern neuroscientist specializing in the ageing brain

LISBETH MARNER
1974–
Neuropathologist who showed age-related white matter changes

30-SECOND TEXT
Daniel Bor

'Old age ain't no place for sissies', according to Bette Davis, herself sampling H. L. Mencken. Of all the ages of man (and woman), the seventh is the cruellest.

THE PARKINSONIAN BRAIN

the 30-second neuroscience

3-SECOND BRAINWAVE
Parkinson's is a debilitating prevalent disease affecting movement and mood, in which the brain's dopamine-containing neurons degenerate and die. There is no known cause or cure.

3-MINUTE BRAINSTORM
Progress in Parkinson's research is being made on several fronts. The prevention of neuron degeneration through the development of anticell death agents shows some promise, as does the gene therapy involving the use of viral carriers to introduce therapeutic genes into specific brain regions. Stem cell research in animals is another exciting area. Here, the aim is to replace dead and dying dopamine-producing neurons with new cells transplanted into motor centres in the brain.

Among the neurodegenerative diseases, Parkinson's is second only to Alzheimer's in its prevalence, with one per cent of people over 60 and four per cent of those over 80 affected. There is no cure and no known cause. Early signs that can anticipate diagnosis by several years include: loss of sense of smell, insomnia, constipation, depression and a tremor in one thumb. Later, shaking in both arms, slowness of movement, postural instability, rigidity, muscular weakness and stooped posture develop. Finally, about 20 per cent of people with the condition exhibit dementia. Parkinson's involves progressive degeneration of particular neurons which, when healthy, release dopamine in other parts of the brain involved in movement control. This impairs the smooth execution of voluntary movement. Drugs can manage these symptoms and slow the disease's progression. However, not all symptoms are easily explained by this mechanism. In advanced stages, when medication may become ineffective, deep brain stimulation (DBS), involving implanting a brain pacemaker, may be beneficial. A cure, based on stem cell transplantation or gene therapy research, remains a distant hope. Perhaps more immediate benefit resides in recognizing subtle signs that appear years before normal diagnosis. Introducing treatments at very early stages may prevent the progression of the disease.

RELATED BRAINPOWER
See also
NEUROTRANSMITTERS
& RECEPTORS
page 18

3-SECOND BIOGRAPHIES
JAMES PARKINSON
1755–1824
The first to thoroughly document the symptoms of shaking palsy (1817)

JEAN-MARTIN CHARCOT
1825–93
Proposed renaming the disease to honour the name of James Parkinson

ARVID CARLSSON
1923–
Established the crucial role of dopamine depletion in Parkinson's

30-SECOND TEXT
Michael O'Shea

Parkinson's disease gradually blots out the neurons responsible for motor co-ordination, producing the characteristic shaking motion and slow movements in people with the condition.

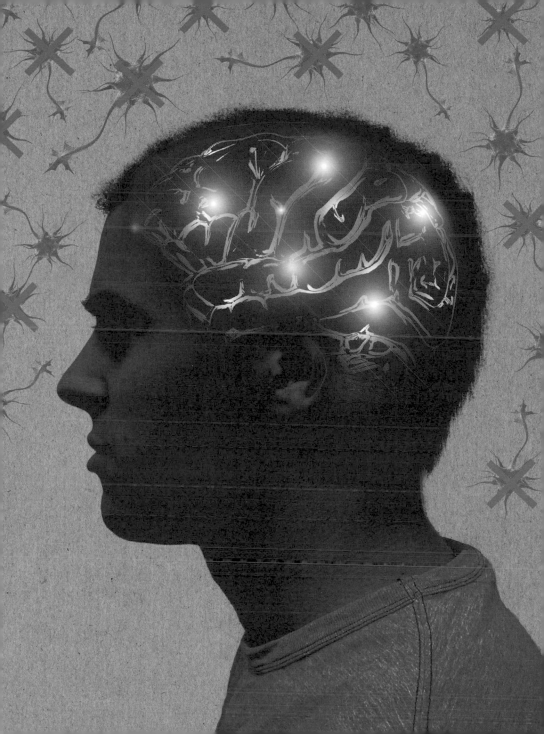

20 August 1913
Born in Hartford, Connecticut

1935
BA from Oberlin College, Ohio

1937
MA in Psychology

1941
PhD in Zoology from University of Chicago under Paul Weiss

1941–46
Fellowships at Harvard

1942
Worked at Yerkes Laboratory of Primate Biology

1942–45
Part of the OSRD Medical Research Unit on Nerve Injuries

1946
Associate Professor of Anatomy at University of Chicago

1949
Diagnosed with TB and sent to Adirondacks for treatment

1952–53
Associate Professor of Psychology at University of Chicago; Section Chief for Neurological Diseases and Blindness at the National Institute of Health

1954
Hixon Professor of Psychobiology at California Institute of Technology

1965
Published first of a series of papers proposing a new theory of mind

1972
California Scientist of the Year

1981
Awarded Nobel Prize (jointly) for Physiology or Medicine 'for his discoveries concerning the functional specialization of the cerebral hemispheres'

1984
Retired but remained Emeritus Professor of Psychobiology at California Institute of Technology

1989
Awarded the National Medal of Science

1991
Lifetime Achievement Award, American Psychological Association

17 April 1994
Died in Pasadena, California

ROGER SPERRY

Ernest Rutherford split the atom in 1917. Forty years later, in work that had much the same seismic effect on his own field, Roger Sperry effectively 'split' the brain, revealing the functions, limitations, co-operation and differences of the two hemispheres and establishing the foundations for brain mapping and new theories of the mind.

Sperry started academic life as an English Literature student and all-round athlete at Oberlin College, Ohio. One of his ancillary courses was Introductory Psychology, and it soon became his overriding interest. After graduating with a BA, he took degrees in psychology and zoology (from the University of Chicago) going on to postdoctoral research at Harvard. After a spell as Associate Professor at Chicago – multitasking in the schools of anatomy, psychology and neurological diseases – he was poached by the California Institute of Technology, where he became Hixon Professor of Psychobiology until his retirement.

Sperry's early research focused on brain circuitry and neural specificity. His elegant experiments indicated that nerves specific to certain activities (seeing or locomotion, for example) could not reroute themselves to reproduce their original function if they were transplanted; at least in these respects the mammalian nervous system seems hardwired and unable to modify or adapt itself. These discoveries provided strong evidence that the development of neural pathways occurred via intricate chemical codes under genetic control, a foundational idea within modern development neurobiology.

But it was while on an enforced sabbatical (he had been diagnosed with tuberculosis or TB) that Sperry began to think about the corpus callosum – the bridge that joins two hemispheres of the brain – a structure whose function no one really understood. 'Splitting' the brain by cutting through this structure proved to alleviate the symptoms of epilepsy without apparent impairment. Sperry's work on 'split-brain' patients showed that the two hemispheres worked independently but in co-operation; without the linking crossover system of the corpus callosum, they behaved much like two separate brains in one head. This research spearheaded a string of discoveries about the lateralization of brain function – for example, that language is usually (but not always) lateralized to the left hemisphere. It also led him to speculate that 'split-brain' patients may, in fact, have two separate and simultaneously existing consciousnesses. Sperry's work won him a share in the Nobel Prize for Physiology or Medicine in 1981. It also led him to speculate on a theory of mind (that consciousness is an emergent function of neural activity), which he considered to be his most important contribution to neurobiology.

THE SCHIZOPHRENIC BRAIN

the 30-second neuroscience

3-SECOND BRAINWAVE
Schizophrenia involves perceptual hallucinations, false beliefs and thought insertions. These symptoms may arise from a failure to properly combine prior expectations with new sensory evidence.

3-MINUTE BRAINSTORM
Several genes have now been associated with susceptibility to schizophrenia. One of them, COMT (catechol-O-methyltransferase), is involved in breaking down the neurotransmitter dopamine, involved in prediction-based learning. The identification of such genetic targets may open new avenues for diagnosis and treatment. However, there is no simple link between genes and mental states. Such conditions involve complex networks of genetic, developmental and psycho-social causes.

Approximately 0.7 per cent of the population will experience schizophrenia at some point in life. Contrary to some common beliefs, its main features are not split or multiple personalities, but a combination of 'negative' symptoms, including emotional flattening and lack of motivation, and 'positive' features, including perceptual hallucinations, false beliefs, paranoid delusions and disorganized thinking and speech. 'Thought insertions' – patients experiencing a lack of ownership of their own thoughts – can be among the most distressing symptoms. Although the brain basis of schizophrenia is not well understood, there are some promising theories. The comparator theory proposes that schizophrenic brains have problems distinguishing between self-generated and externally caused sensations. For instance, when we make a movement, our brain predicts the sensory consequences of the movement so that we experience the movement, as self-caused. If these predictions go awry, the brain may falsely attribute control to some external source, leading to a 'delusion of control'. Recently, the theory has been extended to explain perceptual hallucinations, using the idea that perceptions are also based on predictions. This implies that schizophrenics, unlike most, should be able to tickle themselves, which turns out to be true.

RELATED BRAINPOWER
See also
THE BAYESIAN BRAIN
page 50

VOLITION, INTENTION & 'FREE WILL'
page 88

ALIEN HAND SYNDROME
page 112

3-SECOND BIOGRAPHIES
EUGEN BLEULER
1857–1939
Swiss psychiatrist and contemporary of Freud, who coined the term schizophrenia

CHRIS FRITH
1942–
Pioneer in the neuroscience of schizophrenia

30-SECOND TEXT
Anil Seth

Most of us can differentiate between what we do and what comes at us from outside; schizophrenics have problems with this, so feel constantly under ambush from different parts of themselves.

THE MEDITATING BRAIN

the 30-second neuroscience

The essence of meditation is to train oneself to be as aware as possible of as little as possible. Unlike brain training, meditation is increasingly being shown to have profound effects on thought, emotions and the brain. For instance, long-term meditators have a shrunken amygdala, a brain region associated with anxiety or fear, and an enlarged prefrontal cortex, associated with our highest forms of cognitive processing and intelligence. Long-term meditators also appear somewhat protected from dementia, which makes sense given that meditation causes brain regions linked to complex thought and memory to grow instead of shrink. Activity in the prefrontal cortex can also become more efficient through meditation, so that less activity is needed to perform optimally on a given task. In line with this, long-term meditation improves a range of attentional, working memory and spatial processing tasks. Perception also appears altered, with experienced meditators being able to detect fainter stimuli and being less susceptible to certain visual illusions. Meditation even reduces the need for sleep. Due to its stress-reducing properties, meditation is increasingly being used as a clinical tool, relieving symptoms of chronic pain, depression, anxiety, schizophrenia and other conditions.

3-SECOND BRAINWAVE
In the normal population, meditation has a profound effect on the brain, calms emotions, boosts cognition and can help treat a range of mental illnesses.

3-MINUTE BRAINSTORM
One needn't spend years in intensive meditation practice before any benefits are seen. One study found that just four meditation sessions were sufficient to increase working memory capacity. Another showed that five sessions increased performance on an attentional task that involved resolving conflict. At the same time, just five sessions are also needed to lessen a participant's sense of anxiety, anger and tiredness.

RELATED BRAINPOWER
See also
NEUROGENESIS &
NEUROPLASTICITY
page 138

TRAINING THE BRAIN
page 140

THE AGEING BRAIN
page 144

THE SCHIZOPHRENIC BRAIN
page 150

3-SECOND BIOGRAPHIES
JON KABAT-ZINN
1944–
Pioneer of applying meditation to clinical populations

SARA LAZAR
1965–
Carried out important studies linking meditation with brain changes

30-SECOND TEXT
Daniel Bor

Breathe. Shrink that amygdala, enhance that prefrontal cortex. There is no downside to meditation. Om.

RESOURCES

BOOKS

The Brain: A Very Short Introduction
Michael O'Shea
(Oxford University Press, 2005)

The Brain Book
Rita Carter
(Dorling Kindersley, 2009)

*Connectome: How the Brain's Wiring
Makes Us Who We Are*
Sebastian Seung
(Allen Lane, 2012)

*Descartes' Error: Emotion, Reason,
and the Human Brain*
Antonio Damasio
(Vintage, 2006)

*The Ego Tunnel: The Science of the
Mind and the Myth of the Self*
Thomas Metzinger
(Basic Books, 2010)

*The Frog Who Croaked Blue: Synesthesia
and the Mixing of the Senses*
Jamie Ward
(Routledge, 2008)

Incognito: The Secret Lives of the Brain
David Eagleman
(Cannogate, 2012)

Making Up the Mind
Chris Frith
(Wiley-Blackwell, 2007)

The Man Who Mistook His Wife for a Hat
Oliver Sacks
(Picador, 1986)

The Oxford Companion to the Mind
Edited by Richard L. Gregory
(Oxford University Press, 2004)

*Phantoms in the Brain: Probing the
Mysteries of the Human Mind*
V. S. Ramachandran and Sandra Blakeslee
(William Morrow Paperbacks, 1998)

*Pieces of Light: The New Science
of Memory*
Charles Fernyhough
(Profile Books, 2012)

*The Ravenous Brain: How the New
Science of Consciousness Explains
our Insatiable Search for Meaning*
Daniel Bor
(Basic Books, 2012)

The Rough Guide to the Brain
Barry Gibb
(Rough Guides, 2nd edition, 2012)

The Tell-Tale Brain: Unlocking the
Mystery of Human Nature
V. S. Ramachandran
(Windmill Books, 2012)

MAGAZINES

Scientific American Mind
www.scientificamerican.com/sciammind/

WEBSITES

BrainFacts
www.brainfacts.org
A lot of information about the brain and
neuroscience, hosted by the Society for
Neuroscience and partners.

BrainMyths
www.psychologytoday.com/blog/brain-myths
Stories we tell about the brain and mind,
from contributing author Christian Jarrett.

Frontiers in Neuroscience
www.frontiersin.org/neuroscience
An open-access source of the latest research
in neuroscience.

Mo Costandi's blog
www.theguardian.com/science/neurophilosophy
Mo Costandi's neurophilosophy blog at the
Guardian. Mo is a molecular and developmental
neurobiologist turned science writer.

NeuroPod
www.nature.com/neurosci/neuropod/index.html
A podcast all about neuroscience hosted by the
prestigious journal *Nature*.

The Sackler Centre for Consciousness Science
www.sussex.ac.uk/sackler
The website for a prominent research group
in consciousness science.

Scholarpedia
www.scholarpedia.org
A peer-reviewed version of Wikipedia with many
excellent articles about neuroscience.

Society for Neuroscience
www.sfn.org
The Society for Neuroscience is the world's largest
organization of scientists and physicians devoted
to understanding the brain and nervous system.

EDITOR

Anil Seth is Professor of Cognitive and Computational Neuroscience and founding co-director of the Sackler Centre for Consciousness Science at the University of Sussex. He is also an Engineering and Physical Sciences Research Council Leadership Fellow and a Visiting Professor at the University of Amsterdam. He is editor-in-chief of *Frontiers in Consciousness Research*. He has published more than 100 scientific papers and book chapters, has lectured and written widely on neuroscience for a general audience and his research has been covered in a wide range of media, including the *Guardian* and *New Scientist*. Find out more at www.anilseth.com or follow him on Twitter @anilkseth.

CONTRIBUTORS

Tristan Bekinschtein is a biologist and PhD in Neurosciences from the University of Buenos Aires and has done postdoctoral work in Paris and Cambridge. He is currently Wellcome Trust Fellow at the Medical Research Council and University of Cambridge, and has wide interests in cognition and neurophysiology. In the last few years he has been mainly concentrating on describing different states of consciousness such as awake, sleep, sedation and the vegetative state. His new line of work primarily looks at transitions, how we lose consciousness and how we get it back. He publishes articles in top scientific journals, and also collaborates in TV, radio and media projects involving neuroscience and what makes us humans.

Daniel Bor is a neuroscientist based at the Sackler Centre for Consciousness Science, University of Sussex. He previously worked at the University of Cambridge, where he also gained his PhD. He has published research papers in journals, including *Science and Neuron* on a range of areas, including frontal-lobe function, consciousness, intelligence, memory, brain training, savantism and synaesthesia. He is the author of the popular science book on consciousness science *The Ravenous Brain: How the New Science of Consciousness Explains Our Insatiable Search for Meaning* (Basic Books, 2012). He also contributes articles on neuroscience and psychology for various magazines, including *Scientific American Mind*, *New Scientist*, *Slate* and *Wired UK*. For more information, visit his website at www.danielbor.com or follow @DanielBor on Twitter.

Chris Frith is a pioneer in the application of brain imaging to the study of mental processes. He was one the founders of the Wellcome Trust Centre for Neuroimaging at University College London. He studies agency, social cognition and the hallucinations and delusions associated with mental disorders, such as schizophrenia. He was elected a Fellow of the Royal Society in 2000 and a Fellow of the British Academy in 2008. He is the author of *Schizophrenia: A Very Short Introduction* (Oxford University Press, 2003) and *Making up the Mind: How the Brain Creates Our Mental World* (Wiley-Blackwell, 2007).

Christian Jarrett was the editor of *30-Second Psychology*. He's also author of *The Rough Guide to Psychology*, editor of the British Psychological Society's Research Digest blog, staff journalist for the *Psychologist* magazine, a blogger for *Psychology Today* and a regular contributor to 99u.com, the New York-based creativity think tank. His latest book is *Great Myths of the Brain*. Follow him on Twitter @Psych_Writer.

Ryota Kanai is a cognitive neuroscientist at the Sackler Centre for Consciousness Science and School of Psychology at the University of Sussex. His research focus is to understand the neural basis of conscious experience. He investigates bistable perception, metacognition, binding of visual features and time perception using psychophysics, neuroimaging and brain stimulation. His research has been frequently featured in international media, such as the BBC, the *Guardian*, the *New York Times*, *New Scientist* and *Nature*'s podcast.

Michael O'Shea is Professor of Neuroscience and co-director of the Centre for Computational Neuroscience and Robotics at the University of Sussex. He was previously Professor at the University of Geneva, Associate Professor in the Pritzker School of Medicine at the University of Chicago and Assistant Professor at the University of Southern California. He held National Institute of Health and NATO research fellowships at the University of Cambridge and the University of California at Berkeley. He is author of more than 100 scholarly articles and is author of *The Brain: A Very Short Introduction* (Oxford University Press, 2005). His interests include the molecular mechanisms of memory, non-synaptic chemical signalling, biologically inspired robotics, the hard problem of consciousness and collecting antique scientific instruments.

Jamie Ward is a Professor of Cognitive Neuroscience at the University of Sussex. He has degrees from the University of Cambridge and University of Birmingham. Much of his research has been devoted to understanding unusual perceptual experiences, notably synaesthesia (such as in which music may trigger colours), using methods such as brain imaging, EEG and cognitive testing. He is the Editor-in-Chief of *Cognitive Neuroscience* and also the author of leading student textbooks (*The Student's Guide to Cognitive Neuroscience* and *The Student's Guide to Social Neuroscience*).

INDEX

ACKNOWLEDGEMENTS

PICTURE CREDITS

The publisher would like to thank the following individuals and organizations for their kind permission to reproduce the images in this book.

All images from Shutterstock, Inc./www.shutterstock.com and Clipart Images/www.clipart.com unless stated.

Corbis/Alex Gotfryd: 110.
Getty Images/Hulton Archive: 148.
Ivan Hissey: 44.
Library of Congress Prints and Photographs Division Washington, DC: 64.
Public Library of Science: 80.

All reasonable efforts have been made to trace copyright holders and to obtain their permission for the use of copyright material. The publisher apologizes for any errors or omissions in the list above and will gratefully incorporate any corrections in future reprints if notified.

AUTHOR ACKNOWLEDGMENTS

Anil Seth is grateful to the Dr Mortimer and Theresa Sackler Foundation for enabling consciousness research at the University of Sussex, via the Sackler Centre for Consciousness Science.

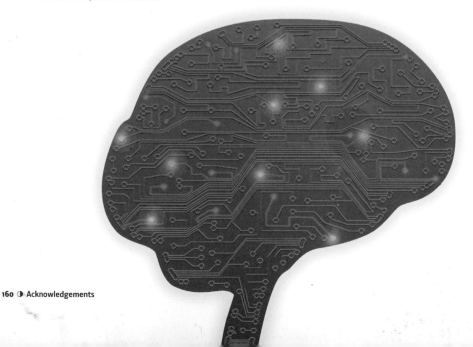